RESIDENTIAL ELECTRICAL WIRING

RESIDENTIAL ELECTRICAL WIRING

A Practical Guide to
Electrical Wiring Practices in Residences

Harry J. Edwards, Jr.

RESTON PUBLISHING COMPANY, INC.
A Prentice-Hall Company
RESTON, VIRGINIA 22090

Library of Congress Cataloging in Publication Data

Edwards, Harry J., Jr.
Residential electrical wiring.

Includes index.
1. Electric wiring, Interior. I. Title.
TK3271.E33 621.319'24 81-15761
ISBN 0-8359-6652-6 AACR2

A Prentice-Hall Company
Reston, Virginia 22090

10 9 8 7 6 5 4 3 2 1

Printed in the United States of America

Interior design/production: Jack Zibulsky
Jacket design: Diane Hoppmann

Preparation of this work emcompassed many hours of typing, proofreading, and preparation of sketches from which draftspersons could prepare the artwork. Even more, it involved many hours of electrical work which gave me the background for the development of this text. It stands as a token of my high regard for electrical craftsmen in the industry and for the professionalism with which they accomplish their work in a safe and acceptable manner. Above all, it represents a sharing of my time by my family, particularly my wife. It is with love and caring, therefore, that I dedicate this work to Fran, my wife, who has always given special meaning to my life. It is only fitting that I salute her now and thank her for being with me in all that I do.

Harry J. Edwards, Jr.

CONTENTS

PREFACE

Much has been written over the years on the subject of electricity, and a significant portion of the material dwells on theory and applied formulae. Other texts, of a more practical nature, treat the subject in a broader form, from the generation of power to its distribution for consumer use, and ultimately cover interior wiring of buildings of all kinds. Even in more restricted treatises, concentration on the specialities of residential wiring is frequently lacking. And yet, this is an area of electrical wiring interest very prevalent among homeowner and expert electrician alike.

It has been my pleasure in this text to attempt to give special treatment to the form of electrical wiring encountered most often by most people, in their homes, and homes of others. Having had much of my dealings with electricity in the residential construction field, I developed a special liking for the form of wiring used in homes, not only in new work, but also in old work, where walls and ceiling framing members are closed in, and getting new wiring into them is quite challenging.

While the book is intentionally uncomplicated, the basis for electrical regulation, the National Electric Code® *, is quite apparent, and I would recommend that a copy be accessible to the reader, so that he can see for himself how the written specifics of each Article referred to lend themselves to the application of electrical work.

*National Electric Code® is a registered trademark of the National Fire Protection Association.

As more and more electrical equipment comes of age in residential use, it would seem logical that homeowners would want a better understanding of what their electrical systems consist of and a little insight into the Code considerations that govern electrical construction prevalent in their homes. Electricians, and electrical contractors too, hopefully, will recognize the special significance I have attached to home wiring, and perhaps glean a pearl or two for their own use.

My offering presumes a knowledge of the basic fundamentals of electricity, so no space is devoted to the laws relating to the details of what electricity is. Rather, the more practical aspect of the application of the electrical wiring system to a home is covered, and in a manner intended to convey the satisfaction I have derived over the years from jobs well done, reflecting thoroughness and satisfaction on the part of the homeowner.

No work can cover everything, but my desire was to be as complete as possible, using actual recollections of job experiences to paint a picture in the first part of the book of the progress of electrical work in a home under construction, and in the latter part of after-the-fact electrical work in older homes. May it please my readers to excuse any oversights or omissions, and look fondly on this work in recognition of the great pleasure I derived from preparing it for your enjoyment and edification.

Permit me at this time to recognize the electrical inspections officials throughout our land, who safeguard our citizens through diligent research and enforcement of the Code, and whose input over the years to the regulatory commissions of the Code Board has assisted in the ongoing updating of the Code Book for use as a primary reference in all matters pertaining to electrical construction.

It is important that my readers recognize that some interpretations of the requirements of the National Electric Code® do vary sometimes from one individual to another. Discussions of the Code herein are based on my understanding and experience in performing electrical work, and while they are felt to be essentially correct, the final authority in interpretation and implementation of Code Articles is the Electrical Inspections Department of the locality in which electrical work is to be performed. Therefore, it is important that matters of interpretation be presented to your local municipality's Electrical Inspections Officials for their ruling, since they are charged with enforcement within their jurisdiction.

Harry J. Edwards, Jr.

INTRODUCTION

Perhaps no part of the home construction system is more intriguing than the electrical wiring portion. Visible receptacles and switches portend hidden wiring that makes its way through walls and ceilings before reaching various outlet points, where it can be observed entering switch and fixture boxes There, the conductors may terminate on the devices they serve, or be jointed with other conductors for circuit continuation. Ranging throughout the house, cables follow unknown paths through dark recesses and inaccessible wall cavities, carrying electrical current to light the night and provide power to equipment. A trip to the attic or basement, or other unfinished portions of the home, may reveal them, observable where they are exposed as they pass through wooden members of the structure, before they disappear once again, dropping down or rising up into partitions to reach their destination.

Homes can be dated according to the kind of electrical installations that run through them. When electricity first came of age, "knob and tube" was in prevalent use in home construction. In this method, single insulated conductors were installed along joists and studs with the use of porcelain insulators, which served the double purpose of supporting the wire and insulating it from the combustible wooden member. Usually attached with nails, they were knob-shaped, and split cross-sectionally at the top with a groove to grasp the wire and secure it when nailed into place. Where the wire passed through structural members, tube-shaped porcelain insulators were employed, affording protection against chafing and contact with the joist or stud. Upon entering fixture, switch, or receptacle boxes, the single conductor was fitted with a flexible, insulating sheath of noncombustible material, installed as a sleeve to guard against contact between wire and box. "Hot" and "neutral" conductors were run separately, and this method pre-dated "cabling," where both conductors are enclosed within the same jacket.

Armored BX cable followed, and represented a significant improvement, in that both "hot" and "neutral" conductors were now enclosed in a common jacket, reducing installation time and adding greater flexibility in the way they were run. BX could be stapled to the sides of studs and joists and penetrate them without additional sleeving. Its armor acted as a ground path and facilitated bonding between metal boxes and service panel. Clamps within boxes served to secure the cable and bond the steel jacket to the steel box. Excellent mechanical protection was afforded by the spiral-wound, flexible steel jacket, and the two or more insulated conductors within were separated from each other by a paper filler, which also prevented movement inside the jacket.

Later periods saw the use of non-metallic cable that did away with the metal jacket. Instead, a fibrous, flame-retardant braid served as the outer covering, resulting in still greater ease of installation and lower material cost. It could be installed with staples, like BX, and could penetrate wooden members without additional sleeving. In addition, it was much easier to cut and split in order to expose the conductors for terminations. Metal boxes with cable clamps served to hold the cable firmly in place. A separate grounding conductor installed within the cable afforded means by which metal boxes and equipment could be bonded back to the grounding electrode conductor at the service panel.

Modern-day non-metallic cable is plastic-jacketed cable. The National Electric Code® refers to it as non-metallic-sheathed cable, and lists its uses and method of installation. Cable staples are employed for support as with earlier non-metallic and BX cable, but plastic switch, receptacle and fixture boxes are available as well. These boxes are nonconductive by nature, and require no bonding to the grounding conductor in the cable. This cable can be seen in widespread use in new homes and older homes where additional wiring is being added. Normally sold in boxed 250-foot coils, it center-feeds from its carton as the electrician makes the installation.

Whichever method you encounter or observe in your own home, the installation was most certainly performed in accordance with the accepted practices of the day and judged to be satisfactory by an electrical inspector representing the local authority with jurisdiction over the area in which the house is located. And, more likely than not, the electrician who did the work took pride in his efforts and represented his craft well by his performance. And so it follows that where electrical changes or additions are contemplated, they should be approached with a good knowledge of practices and requirements to complement the thoroughness of the original work, and they should reflect implementation of updated standards.

There is no place for inferior electrical work. Improperly sized conductors will heat under load, resulting in deterioration of the insulation on the wire and the creation of a potential fire hazard if the wiring heats surrounding wooden members to their ignition points. And poorly terminated conductors create arcing, damaging connected components by heat and oxidation, as the flowing current attempts to bridge the gap created by the poor connection. Because *poor* electrical work constitutes a safety hazard, all electrical work should be undertaken only by qualified personnel. This is not to say that the homeowner cannot perform his own electrical work if he is qualified and his wiring meets local inspection requirements.

In addition to correct methods, adequate safety precautions are to be observed during performance of electrical work. Make sure open portions of an electric circuit are disconnected before attempting to work on it. As basic as this seems, it bears mentioning here, because an electrical shock can be lethal. Electricity is always looking for a path to ground, or to another conductor representing a potential voltage difference. It is all too easy to present yourself inadvertently as the "path" when working around energized parts. Work safely, de-energize equipment before going into it.

Throughout the text, residential voltage is expressed as 115/230-volt. Readers who are more familiar with 110/220-volt or 120/240-volt need to know that 115/230-volt is used nominally here and that such voltage expressions are frequently interchangeable in discussions of residential wiring.

Many methods and techniques applicable to residential wiring will be examined in the following chapters and presented always, hopefully, in accordance with the guidelines of the National Electric Code®, for the enrichment of the reader, that he may feel more comfortable with the electrical features of the typical residence. To this end, illustrations are offered as often as is felt necessary to clarify or outline a topic under discussion. Throughout the book, the overriding theme should come through of sharing experiences of years in the field of doing residential electrical wiring, with readers who are interested in learning more about what they see in the electrical world surrounding them in their homes, and with aspirants to the electrical trade, who may one day find residential wiring work as enjoyable as I have.

Electrical work must always be undertaken by those qualified in the field. For safety sake, inspection procedures should be followed which involve eliciting the approval of the area inspector on what is to be done. Under no circumstances should electrical work be undertaken by the inexperienced, or by anyone who is uncomfortable working with electricity.

The principles and procedures set forth in this book are not intended to encourage novices to proceed into an electrical project without appropriate supervision. Rather, all guidelines and discussions given within these pages are intended for use by those qualified in the electrical field, or by those in training with adequate professional direction to guarantee accurate comprehension of matter presented.

The qualifications of readers vary widely and all non-professionals are admonished to seek competent local directions before beginning to work with electricity, for their own sake and that of others. Safe practices must be the paramount consideration.

The author and publisher precisely disclaim any personal or corporate liability, loss, or risk incurred as a consequence of the use and application either directly or indirectly of any advice or information presented herein.

1

WIRING THE HOME

DEFINING "HOUSEPOWER" REQUIREMENTS

The logical place to start anything is at the beginning. So it is with new home construction, when a plan is developed for the floor space in the house. At that time, room designations are assigned, and thoughts concerning usage of electricity within the areas are formed. The National Electric Code® sets the example in Chapter 9 by establishing a procedure for calculating how much electric power the residence will require. Let us examine the method for deciding upon the electrical service equipment to be used. Electrical requirements are tied to the size of the home and the equipment to be installed, as we see from the example that follows. We will assume a residence of 2,400 square feet, exclusive of carports, porches, or unfinished basement. A washer and dryer are planned for, of course, as is a 4,500-watt electric water heater. An electric range with built-in oven is selected, and heating and air conditioning is to be accomplished by a central heat pump system utilizing a 5-hp hermetic compressor and 20 kw of emergency electric heat as part of the indoor unit. A ½-hp fan delivers air to the conditioned spaces, and a fan the same size handles air through the outdoor unit. From the example that follows, it will be demonstrated how the size of the service equipment is determined, how the size and number of branch circuits are arrived at, and what size wire is to be used for each circuit.*

2,400 sq ft @ 3 watts per sq ft		= 7,200 watts
Two kitchen small appliance circuits @ 1,500 watts each		= 3,000 watts
Clothes washer @ 1,500 watts		= 1,500 watts
Total watts subject to demand factor		11,700 watts
Enter first 3,000 @ 100%	3,000	
Balance at 35%	3,045	
Total net watts after demand factor		6,045 watts
Range		8,000 watts
Dryer		5,000 watts
Water heater		4,500 watts
Total watts before space conditioning load		23,545 watts
Emergency electric heat		20,000 watts
Total watts before motor loads		43,545 watts
Total load thus far, expressed in amperes (43,545/+230)		= 189 amperes
Hermetic motor, in amperes (5 hp @ 230 volts)		= 28 amperes
½ hp 230-volt indoor fan		= 5 amperes
½ hp 230-volt outdoor fan		= 5 amperes
25% of largest motor rating		= 7 amperes
Total load of residence, in amperes		234

This example could very well be typical of many a modern home being constructed today. The steps illustrated draw upon the Code to establish several criteria that deserve further mention.

In Article 220 we are told to allow three watts per square foot of finished living space for general illumination. This allowance covers lighting fixtures and general purpose convenience receptacles. The Article recognizes loads created by small appliances usually found in the kitchen, breakfast room, dining room, and family room areas of the home, instructing us to include two special-purpose circuits at 1,500 watts each. Kitchen refrigeration equipment is covered by these allowances, too. Finally, a clothes washer is provided for by adding in an additional 1,500 watts for its special purpose circuit.

After the wattages for general illumination, kitchen circuits and washer circuit are totaled, the effect of diversity is acknowledged by reducing all wattage above 3,000 up to 120,000 to 35% of itself. Anything over 120,000

---

* Adopted by permission from NFPA 70-1981, National Electrical Code®, copyright © 1980, National Fire Protection Association, Quincy, Mass.

is permitted to be reduced to 25% of itself. The reason for this diversity calculation becomes apparent when we realize that the likelihood of every light fixture and every receptacle being in full use at the same time is quite remote. This makes for a more credible solution to our load calculation problem and establishes load center requirements more in keeping with the maximum demand likely to be encountered.

Referring next to the entry for our electric range, we find the Code once again takes a realistic view of the maximum load that could normally be expected in an operating kitchen. The nameplate rating of the range relates to the total current the appliance would draw with *all* heating elements on high. This includes the oven and surface units together. In actuality, power requirements for the busy range at meal preparation time are substantially lower, owing to the fact that cooking is accomplished at moderate, controlled temperatures, and not on a runaway range with every element top and bottom going full blast.

So Article 220 tells us we are to enter 8,000 watts as the load to be reckoned with on behalf of the combination range and oven. A nameplate rating of 12,000 watts constitutes the maximum size range that may be entered at a simple 8,000 watts. For a nameplate rating of greater than 12,000 watts, an additional 5% must be added to the 8,000-watt entry for each 1,000 watts, or major portion thereof, above 12,000. If the home has double wall-mounted ovens and a separate surface unit, their nameplate ratings may be totaled, and that total treated as one appliance.

An electric clothes dryer is quite commonplace in today's modern home, so our load calculation allows for one. Five thousand watts is noted as the minimum entry we can make for an electric clothes dryer, but we are to use the appliance's nameplate rating if it is larger than 5,000 watts.

Most single-family electric water heaters of the quick-recovery variety make use of two high-capacity elements, one upper and one lower, with

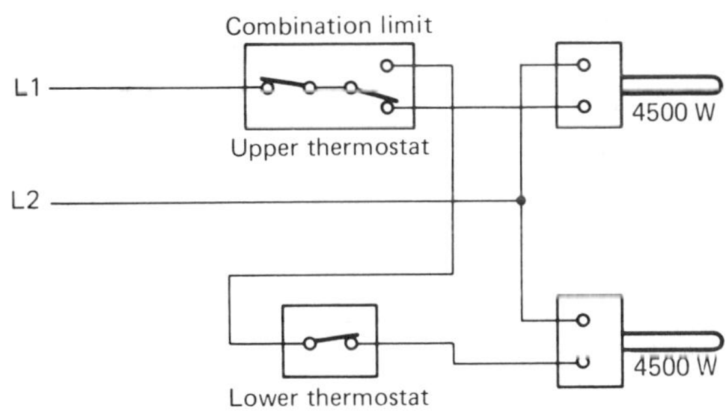

Figure 1: *Water Heater Controls*

interlocking thermostats. Both of the elements may be rated at 4,500 watts, but usually wired is such that only one can operate at a time. The upper thermostat is the master thermostat, transferring power to the lower thermostat only after it is satisfied. Consequently, our load calculation need only concern itself with a single heating element. In Figure 1 we see how the interlock is accomplished. The upper, master, thermostat includes a high limit to shut down both heating elements in the event either of the thermostats should malfunction and cause an overheating condition. The wiring shown is internal to the water heater; only field power wiring needs to be connected by the electrician.

Next in our listing of loads in the calculation is electric heat. Our heat pump utilizes supplemental electric-resistance heat strips in the interior unit to supply heat to the conditioned space if the outdoor unit is incapable of generating enough heat to maintain space conditions. Extremely cold outdoor temperatures coupled with an increasing demand for space heating can create hardship for our heat pump to such an extent that it may not be able to extract sufficient heat from the outdoor air to satisfy interior requirements. In that case, our electric resistance heat strips are automatically energized to supply the necessary heat sometimes coincidental with compressor operation. So an allowance must be made in our load taking their operation into account simultaneously with that of the heat pump compressor motor.

At this point it becomes practical to convert from watts to amperes, because our motor loads are rated in amperes. Watts divided by volts equals amperes. Hence, 230 becomes our divisor, because this is the single-phase voltage with which virtually all homes are served. Derived from a center-tapped transformer, it enters the home as 115/230-volt single phase power, capable of supplying 115- and 230-volt loads. (A subsequent chapter on troubleshooting treats voltage derivation more thoroughly.) Dividing by 230 allows us to state these loads in amperes.

Our outdoor unit contains a hermetic refrigeration motor for pressurization of the refrigerant circuit. In "three piece" heat pumps, this refrigeration motor, or compressor, may be located inside the house, but in any event, it has a nameplate stating its full-load running current in amperes. This is the number we want for load calculation purposes. Also, the outdoor fan motor full-load current must be picked up. It is significant to note that, had our outdoor unit been a "cooling-only" refrigeration unit, the load it would constitute could be dismissed *if* electric resistance heating elements represented the source of heat and constituted a larger load than our cooling compressor. This is the case when the electric heaters and refrigeration compressor do not operate at the same time, as is possible in a heat pump installation where electric heat is supplemental and coincidental. Article 220 makes provision for this under "Noncoincident loads."

The indoor part of our heat pump installation utilizes a fan for delivery of air to the rooms in our home under consideration. This motor load is also picked up and entered with our other motors at this time. It is important to determine the exact space-conditioning equipment to be used, so that electrical load calculations can be accurately performed, based on the operational characteristics of the unit to be installed. But for our purposes, we shall assume coincidental operation of the compressor and electric heaters in extremely cold weather.

We have almost reached the conclusion of our exercise, but one important factor remains concerning motors. Article 430 of the Code instructs us to add in 25% of the load attributable to the largest motor as a final entry under motor loads. This is done to insure sufficient margin of safety to protect against minor discrepancy between nameplate current and actual current of motors, allow for restarting of space conditioning motors during normal cycling, and compensate for drooping power factor as motors load up and unload in response to space temperature demands.

On totaling all the figures, we find a final load of 234 amperes. This figure guides us in our decisions as we select service equipment, service entrance conductors and grounding electrode conductor.

This standardized calculation procedure from the National Electric Code® serves as a tool for the electrical designer, and should be used as the basis for electrical planning of each single-family dwelling. Other loads of significance that require special purpose circuits should be entered as well. They might be well pumps, lathes and planers, welders, dishwashers, infrared bath heat lamps, etc., and must be allowed for to insure ample housepower to facilitate full enjoyment of the electrical system being designed into the home. Local inspection officers are generally happy to be of assistance, should questions arise concerning the Code and its applications. They represent the final authority in the area where the house is to be constructed.

DETERMINING BRANCH CIRCUITS AND PANELBOARD

Before looking into the selection of a panelboard to be used as our service equipment, our attention should be directed toward making a determination of the number of circuits our house will require, since this enters into the consideration of how many poles the panelboard should have.

15-ampere circuits are sufficient for our general illumination circuits, covering lighting fixtures and general purpose receptacles. Referring back to our example, we note that 7,200 watts represented the power requirement based on square footage, and this requirement will be met with 115-volt circuits. The Code example in Chapter 9 instructs us to divide the 7,200-watt requirement by our 115 volts to arrive at amperes. This gives an answer of 62.6 amperes, and our decision to use 15-ampere circuits leads us to a need for five 15-ampere circuits.

Our two kitchen circuits and our clothes washer circuit must be 20 amperes, as required by Article 220 of the Code. So we note a requirement for three 20-ampere circuits.

As regards our range, we recall that 8,000 watts constituted the load that it contributed. When divided by 230 volts, this yields 34.8 amperes, so we will set up a 40-ampere circuit to serve our range.

The clothes dryer, at 5,000 watts, equals 21.7 amperes, and we select a 30-ampere circuit to serve it.

The water heater will require a 30-ampere circuit also. 4,500 watts divided by 230 volts produces 19.5 amperes, but we cannot load a circuit beyond 80% of its capacity where a continuous load is involved. The Code in Article 100 defines a continuous load as "one where the load may be expected to continue for three hours or more."* In the case of a water heater, continuous operation could very well occur, e.g., where there was a heavy, sustained demand for hot water. Article 220 points out that our circuit must be rated at 125% of the load.

Next in our example calculation is a load of significant proportions: that of the electric heating elements in our indoor unit for space heating. After dividing 20,000 by 230, we arrive at an answer of 86.9 amperes; but heating is a continuous load, so we multiply by 125%, to raise our circuit rating requirements to 108.6 amperes. So here we would select a 110-ampere circuit, even though actual current will be below 100 amps.

The outdoor unit of our heat pump system contains both the compressor and outdoor fan, for movement of air across the refrigerant coils in the unit to promote heat exchange. Since the compressor draws 28 amperes, and the fan draws 5, our nameplate rating states 33 amperes as the full load ampere requirements for the unit. The nameplate will give the maximum size of the circuit protective device also. This usually parallels the information given in Article 430 relative to maximum settings of motor branch circuit protective devices, but the nameplate rating is the one to be used, because the manufacturer has tested and rated his piece of equipment, and the manufacturer's rating must be adhered to for proper equipment protection.

Frequently, the nameplate mentions maximum *fuse* size, and most inspection officials agree that this calls for a fuse, rather than breaker, some-

*Reprinted by permission from NFPA 70-1981, National Electric Code®, copyright © 1980, National Fire Protection Association, Quincy, Mass.

where in the circuit, to effect the protection required by the manufacturer. In practice, the circuit may originate from a breaker panel, but is routed through a fusible disconnect located at the piece of equipment as required to facilitate "fusing."

But our consideration here is with branch circuits, and that relates to protective devices in the panelboard. Our choice here will be for 50-ampere circuit, with #6 copper wiring. This gives us a wire size as large or larger than the unit's nameplate wire size requirement, and in-line with requirements for applying 125% factor for continuous loads.

Probably the unit nameplate's maximum fuse size listing would be on the order of 50 amperes also, and if this were the case, we could set a 60-ampere fusible disconnect at the piece of equipment and install 50-ampere time-delay fuses. An extra level of protection is provided in this case with redundant protection located right at the outdoor unit, because no current would be allowed to pass greater than that of the in-line 50-ampere fuses. In this instance, our initial circuit protection in the form of the 50-ampere breaker would protect the wiring against shorts or grounds, and the 50-ampere time delay fuses would protect the unit against an overcurrent condition in excess of their rating.

And so, only our ½-hp indoor fan motor is left. But we have already provided for a 110-ampere circuit to our indoor unit where the electric heat strips are located, and this same circuit will serve the contained fan as well. A manufacturer's nameplate would appear on this unit also, and it would be necessary to guarantee that we met the minimum in wire size and did not exceed the maximum in circuit protection size. But for this consideration, we can assume that the 110-ampere circuit is sufficient.

Now that we have sized our circuits, we can select our panelboards and the breakers to go in them. Selection is made on the basis of available equipment of sufficient sizes to meet our requirements for 234 amperes. Panelboards usually are available in 100-, 125-, 150-, and 200-ampere sizes for residential 115/230-volt single phase service. And we need to be cognizant of a limitation in Article 384 applicable to residential lighting and appliance branch circuit panelboards used as service equipment that restricts such panelboard size to 200 amperes. So our selection here will be for two panelboards, one of 100-ampere capacity, and the other of 200-ampere capacity, each with a main breaker for panelboard protection. Figure 2 illustrates our service equipment.

This 300-ampere services will satisfy the requirement we have for 234 ampere's of housepower, and allow ample spare breaker space for electrical system growth, should it be desired later to add a circuit or two. The view is of the installed panelboards, service entrance conductors, and grounding electrode conductors. No breakers are in place, and we can observe the construction of the current-carrying members within the panelboards. Touchdown points for the "hot" connection to feed our breakers are extended into the midsection of the panelboard on an alternate basis. When the breakers are inserted into position, they automatically make contact with these horizontal bars. This staggering of connection points for our circuit breakers insures balancing of loads with respect to which "hot" service conductor is called upon to handle the connected circuit, and provides for 230-volt power wherever a 2-pole breaker is inserted.

Some manufacturers' circuit breakers above 70 amperes use up a total of four single-pole spaces, owing to the size of the current-carrying parts of the larger breakers.

In dealing with "loading" our panelboards, let us start with the smaller of the two. We need the additional pole spaces in our larger panel for the majority of our circuits and for future loads. By putting the range, water heater, and dryer in the smaller panel, we use half of its spaces and load it to less than 80% of its panelboard rating. One 2-pole 40-ampere and two 2-pole 30-ampere breakers will accomplish this. Actual consumption of the 5,000-watt dryer and the 4,500-watt water heater is 22 amperes and 20 amperes, respectively; and for the maximum 8,000 watts likely to be required for the range, consumption is 35 amperes. This adds to a total of 77 amperes actual load in our 100-ampere panelboard. Of course, the two 30-ampere circuit breakers serve the dryer and water heater, and the one 40-ampere breaker will serve the range.

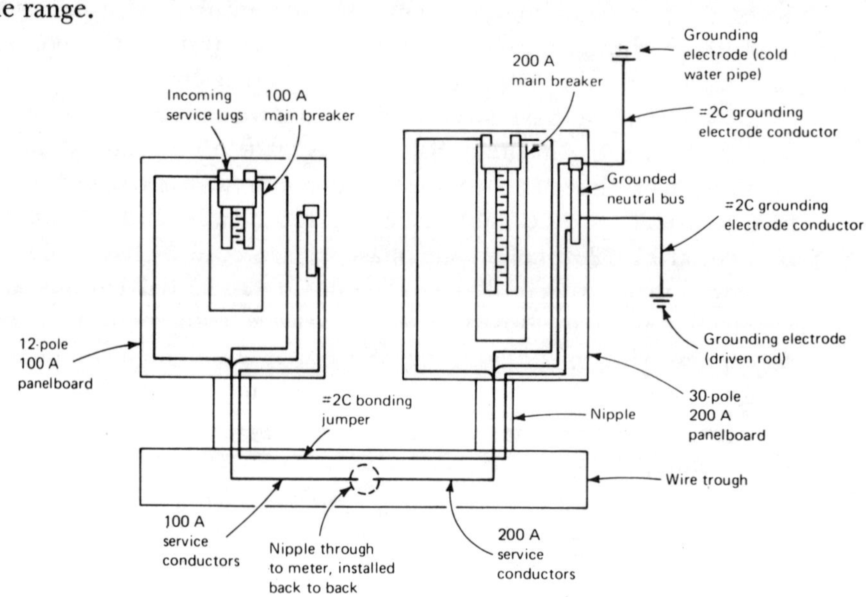

Figure 2: *Service Equipment*

Our larger panelboard will handle the balance of the loads. One 2-pole 110-ampere breaker will serve the indoor unit with 87 amperes of electric heat and fan motor. Another 2-pole 50-ampere breaker will carry the outdoor unit with its 28-ampere compressor and 5-ampere fan motors. Then the balance of our loads, all of which were subject to our feeder demand factoring, can be served through three single-pole 20-ampere and five single pole 15-ampere breakers for an additional 26 ampere load. (6,045 watts ÷ 230 volts). Add another 6 or 7 amperes for the "largest motor" safety factor as set forth in the example, and we arrive at a load in the 200-ampere panelboard of about 157 amperes, representing a load in this panelboard, too, of less than 80% panelboard rating.

Now, if you add the 77-ampere load in our smaller panelboard to the 157-ampere load in our larger panelboard, you see how we distributed and connected our original 234-ampere load between the two panelboards. In so doing, we used a total of 16 pole spaces in the larger panelboard, leaving 14 spare.

Branch circuit wiring will enter the top of the two panelboards through connectors installed through the knockouts. "Hot" wires of the branch circuit cables will terminate on the terminals of their respective breakers, and neutral and grounding conductors will terminate on the grounded neutral bus.

It is worthwhile at this point to mention that ground fault protection, with either a ground fault interrupter circuit breaker or individual ground fault interrupter receptacles, is required in the bathrooms, outside, and in garages for convenience receptacle outlets located in these areas. If the planned layout of the house is such that receptacles in these areas are served by the same circuit, a ground fault interrupter circuit breaker would probably represent the most economical approach to meeting this Code requirement.

In this two-panel installation, we unload off of the power company's meter socket with two sets of incoming service conductors. The length of their run is minimal, because our panels and meter socket are back to back. Article 310 of the Code gives us size requirements for our service conductors, as follows. For the 100-ampere panelboard, we may use #4 copper, or #2 aluminum, conductors. For the 200-ampere panelboard, we may use 2/0 copper or 4/0 aluminum. Our grounding electrode conductor, the wire that connects the neutral bus in each panelboard to the cold water pipe and driven rod, has its size governed by Article 250 of the Code, which contains a table, listing the grounding electrode conductor size based on the size of the largest "hot" conductor or equivalent size of paralleled conductors. Note that the smaller panelboard, with its #4 copper or #2 aluminum energized conductor would require bonding to the larger panelboard with a #2 copper bonding jumper, the same size wire used as our grounding electrode conductors for the 300-ampere service. Paralleling of the 100-ampere and 200-ampere service conductors produces a requirement for grounding electrode conductor sizing suitable for a 300-ampere service.

At this point, we have selected the service equipment to handle the power requirements of the contemplated house, determined the number and sizes of circuits, chosen service entrance conductors and grounding electrode conductors, and selected breakers for each of the circuits. Next we will consider the wire sizes for the circuits we will be installing.

Up to this point, we have seen the effects of diversity in our considerations of housepower requirements in the plannning of our electrical installation. It is realistic and practical to plan projected usage on the basis of actual usage that has evolved out of experience and family energy demand patterns. Public utilities have been doing this for years; if *every* connected load in a utilities service area were energized at once, capacities would soon become strained, and might fail. Our circuit conductor sizing, however, will match or exceed the rating of its protective device—in this case, circuit breakers in our panelboards. Final overcurrent protection of our branch circuits is never to be larger than the capacity of the wire used as the conductor.

SELECTING WIRE SIZES

Focusing our attention now to selection of proper conductors for use in our branch circuits, we refer first to Article 336 covering nonmetallic-sheathed cable, and find that the current-carrying capacity, or *ampacity*, is that given in Table 310-16 of the Code for 60°C insulated conductors. We will choose copper wiring, since it is listed first in the table, so that referencing is easier. The following ampacities are given for copper wire sizes in cable form: #14 gauge, 15 amperes; #12 gauge, 20 amperes; #10 gauge, 30 amperes; #8 gauge, 40 amperes; #6 gauge, 55 amperes; #4 gauge, 70 amperes; #3 gauge, 85 amperes; #2 gauge, 95 amperes; #1 gauge, 110 amperes. From this selection, we can choose the proper size wiring for each of our branch circuits required for our hypothetical house.

Our largest circuit utilizes 110-ampere wire, or #1 gauge. This would be installed to serve the indoor section of our heat pump, and connected to the 2-pole 110-ampere breaker in the larger of our panelboards.

The next largest wire size requirement would be necessitated by the outdoor unit portion of the heat pump. Breaker size of 50 amperes is utilized to provide protection from short circuits and accidental grounding, and final overcurrent protection of the circuit is furnished by the 50-ampere fuses in the local disconnect. So wire size may be chosen based on the 50-ampere protection afforded by our fusible disconnect. #6 is called for here.

Then comes our range, served from the 2-pole 40-ampere breaker, and we quickly see the need for minimum size #8 wire.

The water heater and dryer each obtain power through 2-pole 30-ampere circuit breakers, so this calls for #10.

Our 20-ampere clothes washer and kitchen circuits indicate a need for #12 wire, and the general purpose illumination and receptacle circuits of 15-ampere capacity will have #14 wire.

Notice that our selections of wire size rely upon current flow, expressed in amperes, to determine conductor magnitude. *Amperes* relates to the speed with which power, in watts, is transmitted through the conductor to the connected load. The applied voltage, or electrical pressure, has the effect of lowering amperes, or speed, as it (voltage) is raised. All of the wiring insulation in non-metallic cable is rated at least 600 volts, so our selection process seeks to select suitable conductor size to handle the ampere requirements of any given circuit. Our circuits are connected within the panelboard to either single-pole or 2-pole breakers for either 115-volt or 230-volt potential, respectively. (Potential may be taken to be a kind of electrical "pressure".) Served by 115 volts, 1,500 watts would create a load of 13 amperes, but if this same 1,500-watt load could be alternately connected so that 230 volts be applied to serve it, our amperage would drop to 6.5. And this would permit the use of #14 wire instead of #12. That's why the larger loads in your home are connected for the higher voltage: to maintain minimum size wire and equipment requirements in the interest of economy. The versatility of the two-voltage single-phase residential power service makes either of the two voltages available within the panelboard by merely installing the proper breaker and correctly terminating the wires.

The 115-volt circuit will have the black conductor connected to the screw terminal of the circuit breaker, and the white conductor terminated on the grounded neutral bus bar. But 230-volt circuits will have both the black and the white conductors connected to screw terminals of a 2-pole breaker. Both the circuits, however, will have a third, grounding conductor, terminated on the grounded neutral bus bar also, except that most localities require separate neutral and ground bus bars, where panels are sub-fed with SER cable.

As an aside to our consideration of wire size here, it is appropriate to mention that 2-pole breakers must utilize handle ties, so that "common tripping" occurs. Never use two separate single-pole breakers for 230-volt service, unless special manufacturers' handle ties are installed, so that an overload or short in one leg will have the effect of opening both legs of the circuit.

LOCATING OUTLETS

Continuing through the planning stage, we will now approach the subject of selecting the locations of outlets within the home, so that they may be shown on our house plans for the electricians making the installation. Article 100 of the Code defines an outlet as "a point on the wiring system at which current is taken to supply utilization equipment."* In practice, this covers any point where conductors are bared for terminations on devices or equipment, or for jointing with other conductors. So when we plan our outlets, we are talking about planning the locations of switches, receptacles, fixtures, and equipment.

Highlighted now are the requirements of Article 210, because they will regulate our thinking in the area of outlet planning.

Receptacles are to be spaced so that no point along the wall is more than 6 feet distant from a receptacle. This calls for the placement of a receptacle every 12 feet along the wall. Any wall space at least 2 feet wide must be considered as wall space that requires a receptacle. This requirement is quite genuine, because its intention is to prevent stringing of extension cords or lamp cords across doorways. Consider a wall with two doors in it and a 3-foot space between the doors. It is not unusual to find a table with a lamp on it placed between the doors, and were a receptacle not provided in this small wall space, the lamp cord would have to be stretched across a doorway to reach a receptacle. Or an extension cord might be stapled around the door moulding, so that the lamp could be plugged in. This would constitute an illegal use of an extension cord.

Counter spaces in kitchen and dining areas are to have a receptacle over them if they are 12 inches wide or more. And if a counter space is separated by a sink, range top, or refrigerator, an additional receptacle is to be provided over the additional counter space; and here we are to keep in mind Article 220, which reminds us that kitchen receptacles are part of our small appliance circuit, rated at 20 amperes and wired with #12 wire. Additionally, two such circuits are to serve kitchen receptacles, and either or both of the circuits are to be extended to dining room, pantry, and breakfast room receptacles, as well. In planning wire routing, we will remain cognizant of the

*Reprinted by permission from NFPA 70-1981, National Electric Code®, copyright © 1980, National Fire Protection Association, Quincy, Mass.

requirements to circuit these small appliance receptacles separately from circuiting the rest of the house, and to use the heavier gauge wire.

Of course, a receptacle is required for our clothes washer, and we recall that the washer is on its own independent circuit rated at 20 amperes. The receptacle must be within 6 feet of the washer location.

In the bathroom, a receptacle is required adjacent to the basin. One receptacle is also mandatory outside the house and in attached garages. These receptacles just mentioned will have to have ground-fault protection. If it is convenient to put them on the same circuit, our breaker could be of the ground-fault interrupter type, and all receptacles on its circuit would be protected by virtue of their connection to it. The outside receptacle may be excused from the ground-fault protection requirement by local inspections officials if it is not accessible by direct grade access, as might be the case for one located on an upstairs porch. But in our house, our location in the outside wall of the house for use at ground level definitely calls for ground-fault protection.

Article 210 continues with further instructions to provide lighting outlets, e.g., light fixtures, with switch control for every habitable room in the house, plus bathrooms, hallways, stairways, attached garages, and at outdoor entrances. Then, in reference to other areas of the house not so habitable, but which must be entered on occasion, we are told to provide a lighting outlet for crawl spaces, utility rooms, basements, and attics if these areas are used for storage or to accommodate mechanical equipment that may require service, such as a heating plant. However, these lighting outlets do not have to be wall-switch controlled, and we are allowed the option of using pull-chain-operated fixtures.

An exception in this same Article allows us some latitude in fixture requirements by permitting one or more wall receptacles to be wall-switch controlled in habitable rooms other than baths and kitchens in lieu of placing permanent lighting fixtures there. This means that in bedrooms, studies, living rooms, dining rooms, dens, family rooms, recreation rooms, parlors, libraries, or similar rooms, we can dispense with the use of permanent wall- or ceiling-mounted light fixtures if we so desire, and instead provide a wall switch connected to control a receptacle into which a lamp could be plugged. This would enable us to turn on a light at the doorway before entering a room, as required by the Code, even if we found the use of permanent ceiling-type light fixtures objectionable. And where the room has multiple entry points, we can install a switch at each door to control the same receptacle or light fixture from more than one location, so that regardless of the direction from which we enter the room, we can light it first at the doorway, so that we are not groping for a lamp in a darkened room.

Article 210 also mentions a need for a receptacle outlet wherever flexible cords are used. So in planning and locating our receptacle outlets, we should keep in mind placement of cord-connected equipment, such as sewing machines, televisions, window air conditioners, dressing-table lamps, and the like, which, by our selection of their location, establish a need for a receptacle there to facilitate connection without the use of extension cords.

It is quite often desirable to install lights in closets where darkness would render the closet less serviceable. Article 410 has a word of caution concerning fixtures in clothes closets. Pendants are generally prohibited in most areas, owing to an interpretation of the Article that rules out the use of exposed bulbs. Some localities may permit the use of fixtures with bulbs covered by globes if they are installed inside the closest on the wall over the door with at least 18 inches of horizontal clearance from shelves where items may be stored. Or they may be permitted if installed on the ceiling of the closet with at least 18 inches of horizontal clearance to floor level, so no storage of material may occur below the fixture, where items could be stacked on shelves and possibly come in contact with the fixture.

The preference, by and large, and in many localities the only acceptable fixture for use in closets, is for one that is recessed and that contains a solid lens flush with the ceiling of the closet. The Code considers these fixtures to be outside of the closet; therefore their use is universally acceptable to enforcement authorities. Both pull-chain- and wall-switch-controlled models of recessed closet fixtures are available.

Equipment outlets are not to be overlooked. Our indoor portion of the heat pump will make use of a fusible disconnect located at the piece of equipment. This is a safety switch in an enclosure with an externally operative handle and internal fuse holders, or a fusible block with pull-handle in an enclosure. It will be mounted adjacent to or on the piece of equipment and be readily accessible to personnel who may have need to de-energize the unit for maintenance or repair. The outdoor unit will be treated likewise, but its enclosure will be weatherproof.

Our water heater will be permanently wired in through flexible conduit, extending from the wiring compartment on the water heater up to a junction box (perhaps in the wooden floor joists above, if our home has a basement), where conversion to non-metallic sheathed cable would take place for the home run to the panelboard. More likely than not, the water heater will be placed in the utility area near washer and dryer, and flexible conduit will run to an adjacent wall-mounted junction box. Of course, the dryer will utilize a special-purpose dryer receptacle, either flush or surface mounted, and similarly for our range. Sometimes, built-in ranges feature flexible conduit leads, which require a junction box in back of the unit.

Our discussion of outlets, for all practical purposes, deals with minimum Code requirements, and additional receptacle and lighting outlets are the decision of the builder. Generally, if in doubt whether an area needs an additional receptacle or two, add them in: it is harder to do so later, when walls are already up.

With outlet locations decided, some thought needs now to be given to routing of our wiring. At the same time we plan where we want our outlets, we can also look to the paths we may use for making our cable-pulls later on during the installation stage, with an eye toward total footages of various cables required. And planning roughly how it can be accomplished will aid immensely in selection of outlet box sizes.

BOX SIZING AND CABLE ROUTING

Preparation of the bill of materials cannot go forth until we examine our outlet locations in light of our planned routing of the cables. Certain size boxes for receptacles, switches, or fixtures are required, based on the number of wires the box will have to accommodate. Article 370 of the Code points out that the cubic inch volumetric area within a box is important to prevent crowding of wiring and joints within. There must be so much free space within each box, based on the number and size of conductors in the box. For each #14 conductor, there must be 2 cubic inches, for each #12 conductor, 2.25 cubic inches, and for each #10 conductor, 2.5 cubic inches, of free space. Let us examine some partial routing diagrams, and see how the application of this rule influences box size. Allowances must be made for cable clamps, fixture studs, strap-mounted devices (switches, receptacles) to be attached to the box, and grounding conductors. They all take up room, and provisions are to be made for accommodating them without crowding. Each device counts as one conductor, and one or more cable clamps, fixture studs, or grounding conductors count as one conductor, also, in considering box sizing.

Cable routing diagrams, such as Figure 3, can be quite helpful both in making sure that no outlet is overlooked and for future reference, should later additions or modifications to the electrical system be necessary.

Figure 3 permits us to picture boxes and cables and their relation to one another. Note that no dining room or kitchen receptacles are shown on this circuit, because they must be circuited separately with heavier wire.

Receptables with just two cables entering the box can be sized on the basis of four conductors plus one for grounding conductors entering the box,

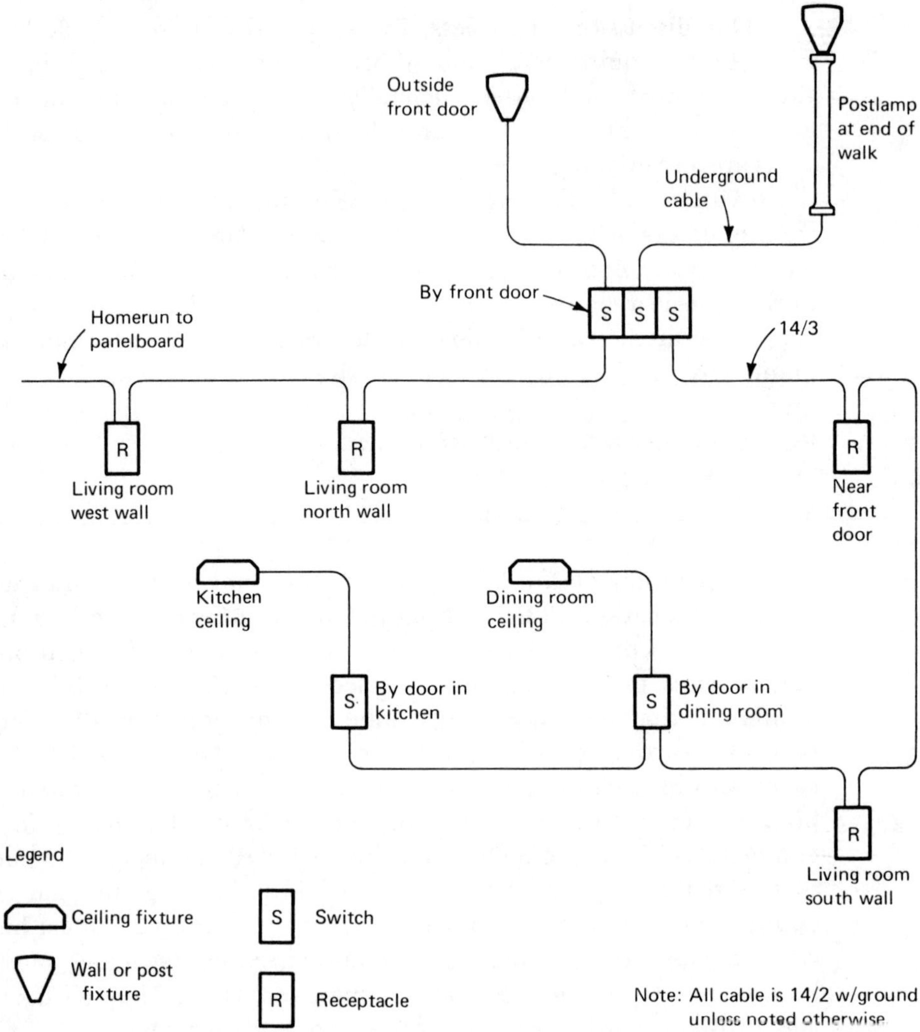

Figure 3: *Cable Routing Diagram for Typical First Floor Light Circuit*

one for the receptacle device to be installed later, and one more for cable clamps, which we will assume. This gives us a total count of seven, and we are to allow 2 cubic inches per conductor. Thus, the box must offer at least 14 cubic inches. Referring to Article 370, we note that a standard 3-inch x 2-inch switch box 2¾ inches deep will meet that requirement. Our switch box for the dining room light has a third cable, though, and we will require an additional 4 cubic inches. So we choose a 3-inch x 2-inch switch box 3½ inches deep, which offers 18 cubic inches, more than enough.

The ceiling fixture boxes have cable clamps and only one cable. No mounted devices attach to the box, as the fixture mounts flush with the ceiling and the edge of the box. Grounding conductor, however, is of course present, and we will need a box that offers at least 8 cubic inches, 2 cubic inches each for our two conductors, another 2 for the grounding conductor, and 2 more for the cable clamps. A 4-inch round or octagonal box 1¼ or 1½ inches deep will do nicely. This same size box will handle our light outside the front door.

Now let us look at the three switches located by the front door. Our plan must provide for a 3-gang box or three single-gang gangable switch boxes. There are three cables with two conductors each, and a fourth cable with three conductors. The piece of 14/3 is used to save doubling back in our rough-in path. The receptacle by the door is to be controlled, in lieu of a permanent fixture in the living room. But we also want to extend our constant hot conductor and our neutral through the receptacle box by the door on to remaining outlets on the circuit, and the use of a piece of 14/3 here will accomplish that for us. The third conductor will be connected to the hot terminal of the receptacle and be controlled by the switch.

With three 2-wire cables and one 3-wire cable coming into our 3-gang switch box, we have established a need for 18 cubic inches, based on conductors alone. An allowance for the three switches adds 6 more cubic inches, grounding conductors add 2 more cubic inches, and cable clamps add another 2, for a total of 28 cubic inches required within the 3-gang assembly. Gangable 3-inch x 2-inch switch boxes 2½ inches deep offer 12½ cubic inches each, so all together there are 37½ cubic inches of free space, ample for our needs here.

Establishment of our requirements for boxes of various sizes at different locations can aid significantly in the preparation of a good materials list.

SETTING BOXES

The first order of business, upon arriving at the new home to begin our work, is to locate and set the switch and receptacle boxes and set fixture boxes and recessed fixture housings. Once in place, our boxes establish a definite point toward which we work, as we bore holes and pull cable. Failing to set all boxes at the outset can result in incorrect routing and wasted wire.

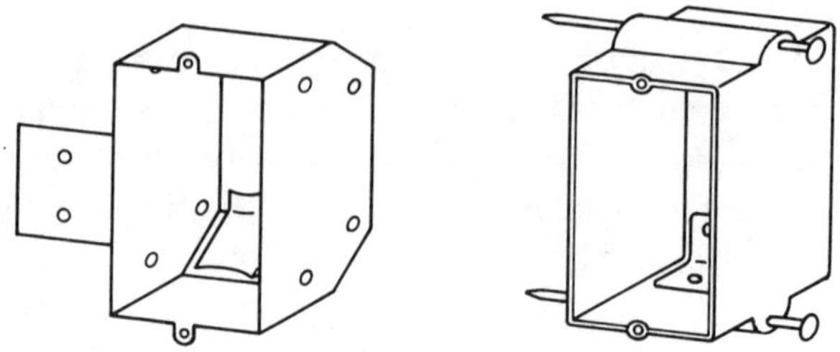

Figure 4: *New Work Rough-in Boxes for Switches and Receptacles*

Figure 4 illustrates two typical kinds of boxes, we might use: beveled-corner 3-inch x 2-inch metal switch box with cable clamps and pryouts, to permit insertion of the cable; and a non-metallic box with captive nails, knockouts for cable entry, and cable clamps also. The first box has holes drilled in the face of the angled bracket to permit nailing onto the face of a stud. Both boxes offer ease of installation with a hammer, attaching directly to the wooden stud.

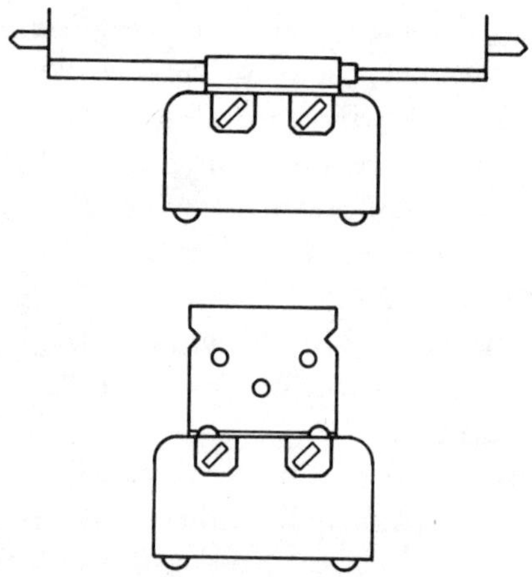

Figure 5: *Rough-in Boxes for Fixture Outlets*

In Figure 5 we have two typical ceiling-fixture octagon boxes. The first utilizes a telescoping bar hanger with cleats on the ends to drive into ceiling joists. Holes in the drive pads permit reinforcing the attachment with nails. The second box uses an L-bracket, also with cleats to facilitate quick attachment, and nail holes to permit reinforcement. Both boxes have pryouts for cable entry and clamps internally for snugging down on the cable when it is inside the box.

Wall receptacles are generally positioned 15 inches above finished floor to the center of the receptacle, except over counters, where 44 inches to the center is usually the case. Switches are generally located 52 inches above finished floor. Bathroom receptacles and clothes washer receptacles most often are 44 inches above the floor also. The Code does not tie down elevations of switches and receptacles because flexibility is necessary for special circumstances, e.g., when the owner prefers his wall outlets to be positioned in the baseboard rather than the wall proper, or when predominant patterns in specially paneled rooms dictate the need to raise or lower a light switch to blend in with the overall architectural appearance. Needless to say, however, the elevation selected should render the switch or receptacle serviceable.

Once elevations are decided on in each room, a rule should be used to mark the location at the proper height on the framing member to which the box is to be attached. After all markings are in place, the boxes should be nailed on. In situations where pryout accessibility is questionable after box installation, pryouts should be removed before mounting. The cable routing diagram may be referred to to insure removal of only those pryouts that are appropriate.

Boxes should be set back sufficiently to prevent protrusion beyond the finished wall surface, which would cause receptacle and cover plate to protrude from the surface as well. Our objective is to have cover plates of switches and receptacles come flush with the wall surface. To achieve this, the finished wall thickness in each room should be ascertained before mounting the boxes, thereby enabling them to be brought forward just enough. Article 370 allows as much as ¼-inch setback to the edge of the box from finished walls and ceilings *if* such wall or ceiling construction is of noncombustible material. Otherwise, no such setback is permitted: the box must be even with the wall surface. Ideally, receptacles or switch plaster ears will draw down to the finished wall surface when the device is installed later on. This is illustrated in Figure 6, where the mounting strap for the receptacle

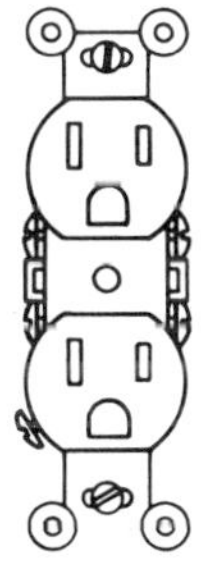

Figure 6: *Receptacle*

features T-shaped tab ends. These tab ends overlap the vertical dimensions of the rough-in box, so that the wall finisher's cutouts, if done properly, will leave the wall surface close around the box for support and leveling of the device as its mounting screws draw it down into place.

When installing switch boxes at doorways, precaution should be taken that proximity of the box to the door opening will not prevent installation of wood trim around the door frame. It is generally advisable to add 2-inch x 4-inch spacers to stand the box off enough to forestall the necessity for later relocation of the switch box. Invariably, 2-inch x 4-inch scrap material is created in the job, and it can be used to mount the box and provide ample clearance for future finishing work. (See Figure 7.) Care should be taken to locate the switch on the strike side of the door, so that the door, when opened, does not hide it.

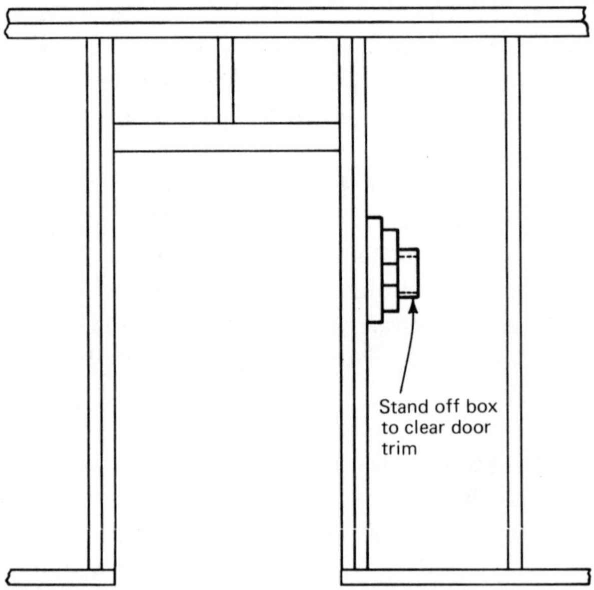

Figure 7: *Switchbox on Framing at Doorway*

Ceiling fixtures, such as dining room chandeliers, kitchen fixtures, bath fixtures, etc., are normally centered in the room. Installing string lines corner-to-corner diagonally will quickly establish the center of the room at the point where the lines cross; or careful measurements of the width and length of the room can be used when plotted to the center point of the ceiling joists. An extremely heavy fixture may require special rigging for hanging. The Code, in Article 410, advises that any fixture in excess of 50 pounds must be supported independently of the outlet box. Generally, however, care should be taken to see that fixture outlet boxes are securely installed to forestall difficulties that could arise later from a box that is improperly fastened.

The edge of the ceiling outlet box should be even with the finished ceiling. During rough-in, as we install the boxes, we should determine ultimate ceiling thickness and make appropriate allowances. For ½-inch ceiling thickness, the edge of our ceiling outlet box can be permitted to come ½-inch below the edge of the ceiling joist.

Ceiling outlet boxes with adjustable hanger bar can be positioned to either of the two far sides of the bar to achieve centering between two joists. Box bracket screws, when loosened, permit the box to move one way or the other, and when retightened, secure the box firmly in place.

Recessed fixtures should be set at this time also. Article 410 of the Code cautions us to use recessed fixtures that are specifically designed and approved for the purpose. Housings of recessed fixtures must maintain ½-inch of clearance from combustible material and 3 inches of clearance from insulation.

Occasionally, the wiring layout of a house under consideration may require joints and continuation of a circuit through a fixture box, rather than just a single switched leg to the fixture. Where this occurs in recessed fixtures, special precautions must be taken to use only recessed fixtures approved for "feed through" wiring. This term has to do with the accessibility of the joints made between the conductors. They must be accessible at all times, and in recessed fixtures approved for "feed through" wiring, this can usually be attained by dropping the lens and fixture interior, which is removable. The construction of this approved fixture is such that the circuit conductors are set apart from the lamp housing, so as to be protected from frequently encountered high temperatures. Then special fixture wire extends into the main portion of the fixture. Installing a fixture with "feed through" wiring approval noted on it ensures that it will meet approval of local authorities and be serviceable in years to come.

Recessed fixtures are rated according to the largest size lamp that can be installed in them. Consideration needs to be given to how much light may be required in any specific area where recessed fixtures are contemplated. Enough fixtures capable of containing lamps of the desired size should be chosen, because recessed fixtures are roughed in permanently, and are not easily changed, as are those that surface-mount to the ceiling after finish work is done.

Figure 8 shows the recessed housing in place with cables entering the accessible junction box, separated slightly from the fixture proper. In this type of fixture, access would most likely be gained through the top portion

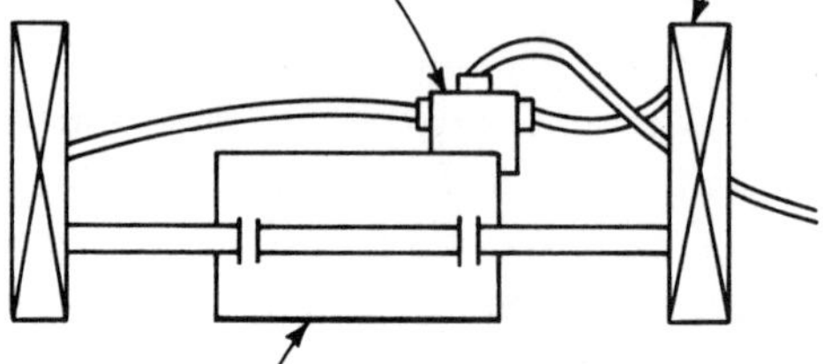

Figure 8: *Recessed Fixture Housing Roughed In*

of the housing, designed to be removable from below once it is permanently in place. The three cables would be for circuit in, circuit continuation, and switched leg. In this installation, one cable is extended to the switch box. Had our hot circuit gone directly to the switch box instead, and continued on from there, then only one cable would have been run to our fixture.

Bath fans represent another type of recessed fixture that must be installed when boxes are set in place. These fans are required when no window is present in the bathroom. Roughing in for the fan entails the installation of ductwork from the fan housing discharge outlet to the weathercap to be mounted on the outside of the house. Generally, the duct is purchased with the fan, even if not made by the fan manufacturer. Mounting of the fan housing is quite similar to that of the recessed fixture just discussed.

If the bath fan unit combines a light and/or heater with the fan itself, then special switches will also probably be purchased with it. Note the rough-in requirements for these switches. Frequently, double-gang switch boxes are required to accommodate the multiple switches associated with being able to energize just the fan, or just the light, or just the fan with heat. Multiconductor cable will be required to accomplish the necessary switching functions. Connections peculiar to combination-fan units will be discussed under a later section.

Through-the-wall exhaust fans, often found in kitchens, should be set at this time, as well. Normally, the fan is removable from the rough-in ring, which contains the splice-box. With the ring in place, cable can be run to it. A switch for control of the fan will surely be desired, and this will require another switch box, located so as to be convenient for fan control.

INSTALLING THE CABLE

Once our boxes are all in place, we can begin boring wooden members for installation of cables. A heavy-duty, geared-down drill motor with fluted, spiral bits designed to pull themselves into the wood works best. Select a bit large enough in diameter to facilitate pulling cable through it, but not so large that it would result in a major displacement of wood out of the stud.

The best route, planned previously, may be horizontally across the studs of the room, from box to box. But check local building codes to determine if there is a restriction against horizontal runs in outside walls. Sometimes there is, because wiring tends to interfere with the proper expansion of

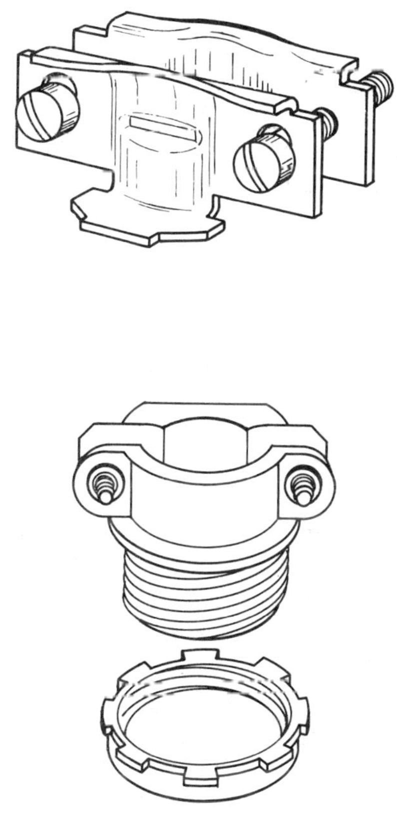

Figure 9: *Cable Connectors*

insulation in outside walls. Where such a restriction exists, rising from below, e.g., from basement or crawl space, may present the best alternative.

Nonmetallic-sheathed cable comes in cardboard cartons, and many feature a center knockout, which permits the cable to feed continuously as the installer pulls it from the carton.

Begin the installation job by pulling the cable through the proper holes, starting at one box, and working toward the next. Do not cut the cable until you have stubbed into the box being roughed into and back-pulled and stapled. Then, with slack taken out, so as not to waste wire, cut the cable, allowing enough to stub into the box at your starting point. Leave at least 6 inches of conductor length in each box.

Article 300 reminds us to bore holes for cables in the approximate center of wooden members. This will help prevent nails from piercing our wiring and do the least possible damage to the stud or joist.

Where cable is run alongside of joists or studs, cable staples must be installed every 4½ feet for support, in accordance with Article 336 of the Code. Drive staples straight in so that they "snug" the cable but do not cut it. When necessary to remove a staple, do not drive a screwdriver under the cable to force the staple to lift out; instead, work an awl in under the staple, *beside* the cable, and raise to loosen. For boxes employing cable clamps, a staple must be installed on the cable within 12 inches of the box; for boxes which do not employ cable clamps, a staple must be installed within 8 inches of the box.

The most popular size staple will handle #14, #12, and #10 cable. Larger sizes are available for larger cables.

If it becomes necessary to use a box that does not include cable clamps, install a cable connector in the ½-inch knockout. Two types, shown in Figure 9, are in abundant use in electrical work. One is two-piece and requires no locknut. After it is inserted into the box opening, the cable is threaded through it into the box, and the screws are tightened until the connector snugs the cable. At the same time, the connector draws together and wedges itself into the opening. The second type uses a locknut to hold it in place while the screws tighten to grasp the cable, as in the first model.

In some instances cable will have to be roughed in without benefit of a box in which to terminate the run. Appliance stub-ups are necessary in kitchens for built-in dishwashers, garbage disposals, trash compactors, food centers, drop-in range tops, or other built-in equipment that won't be set until after cabinets are in place. Take precautions that wiring roughed up through the floor is properly protected from physical damage by finishing trades. "Retrievers" may be used here to good advantage. They are pull wires that are securely fastened to the cut end of our cable, with a round loop on the other end. Once connected, the slack end of the cable can be pushed back through the hole cut for its passage, and the retriever will follow. Leave 3 or 4 inches of the retriever showing, and, using a staple, tack it to the floor. Later, when built-ins are in place and are positioned to cover the unfinished subflooring, our cable will be retrievable through the use of our pull wire, and we can complete our connections. In the interim, the cable will be protected from damage. For cables in walls, it might be advisable to leave them protruding, giving instructions to the people who will be working on the cabinets to pull them through the back of the cabinets, as in the case of a disposal, so we can set a box later on. But those located in floors will definitely suffer from traffic, unless protected, until the equipment they serve is placed immediately over them and final connections can be made.

More than one cable can occupy the same hole through a stud or joist, but good judgment must be exercised in boring large holes. It is not desirable to effectively reduce the cross-sectional integrity of a wooden member at any point. Rather than cut a large hole that might weaken the stud or joist, use multiple smaller holes, bored in the center, and sufficiently spaced so that they do not represent a threat to its strength. Additionally, this will tend to keep cables out of reach of drywall or sheathing nails.

"Cornering" becomes necessary when we reach the intersection of partitions and wish to continue around the room. We must turn the corner and stay within the studs. Framing at corners is designed to afford good edge contact for the finished wall materials. "Boxing" is employed by the framing carpenter at room corners to insure ample support for nailing down wall material edges. Figure 10 shows a plan view of an interior partition abutting

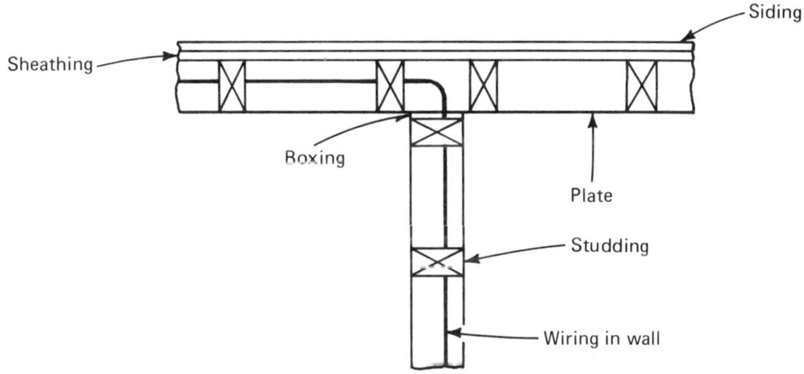

Figure 10: *Plan View of Intersecting Partitions*

an exterior wall. Boxing of studs results in support for each of the two edges of wall material in each of the two rooms. We must penetrate the void in between, turn the corner, and continue our run. Generally, bracing is used at staggered intervals inside the "box" to preserve vertical alignment of stud edges, and on examination, we can usually see where to bore to avoid encountering the bracing, so that we have a clean shot at rounding the corner. Bore a hole, intended to break through into the void, from each direction. Bend the cable in a 90° turn with about a 4-inch radius, and insert it into the hole. Usually the turn can be negotiated, the other hole picked up, and the cable pushed through. If this fails, a pull wire can always be put to use. It may be desirable to go up one bit size, as this helps when cornering by allowing a little more play.

Our water heater and clothes dryer cables will be #10, but a word of caution is necessary in regard to the cable chosen to serve the clothes dryer. Oftentimes the electric motor in the clothes dryer is 115, rather than 230 volts, and, in many localities, that establishes a need for a third insulated conductor. If such is the case where you are performing electrical work, 10/3 with ground will have to be installed for the dryer circuit. A quick check with the inspections people will provide the answer. For the water heater, which is purely 230 volts, 10/2 with ground will be used.

When our open-wall work is completed, it is time to call for a rough-in inspection. The electrical inspector needs to see our in-wall work before finishing materials are applied and our work is concealed. Remember, there can be no inaccessible junction boxes permanently covered by finished surfaces.

TERMINATING CONDUCTORS AT RECEPTACLES

The majority of our receptacles will be installed in outlet boxes in which two cables are present. As we roughed in from box to box, cables were cut and stubbed into each box with a minimum of 6-inches of slack of conductor length. At that time, the outer jacket of the cable would have been removed, up to within ¾ to ½ inch of where the cable entered the box. Our conductors would have been separated, and the insulation would have been removed from the last ¾ of an inch of each wire with a proper insulation-removal tool. This means that, for the most part, we have two black, two white, and two bare wires in each box.

Terminations to the receptacle terminal screws are made on a touch and go basis. Each of the black wires terminates on one of the two gold screws, and each of the white wires terminates on one of the two silver, or "white" screws. A removable factory jumper bonds the two common screws together on each side to facilitate straightaway terminations, one wire on each screw. "Touch and go" refers to the fact that the hot conductor of the incoming circuit touches down on one "hot" terminal screw of the receptacle, and the hot conductor of the outgoing circuit, being continued on to the next outlet box, touches down on the second "hot" terminal screw. The same holds true for the neutral wires.

The green grounding screw is used for terminating the bare grounding conductor of the circuit to the receptacle. Since the grounding conductor is a non-current-carrying wire, it may be bare, or uninsulated. An approved connector, such as a Wire Nut®*, is required when making joints, however, the same as with other circuit wires. We then leave two pigtails, one of which is connected to the receptacle grounding terminal, and the other of which is bonded to metal outlet boxes with a grounding clip, an approved device for fastening the bare wire mechanically and electrically to the metal box itself for purposes of grounding. Or use the green grounding screw if one has been provided by the box manufacturer.

A non-current-carrying wire is one through which no electricity flows in normal operation. It is there to carry off any fault or "stray" current that makes its way to the grounding prod of the plug cap of the appliance being used. It forms a path to ground for any electricity that may appear on the surface of a tool that comes in contact with it through internal insulation breakdown, in order that the person who uses the tool not be subjected to any electrical shock. Since this is an extraordinary situation, the bare grounding wire is not normally conducting electricity, and therefore does not have to be insulated, as mentioned previously. In no circumstances may the bare wire substitute as a current-carrying-wire. Metal boxes, because they could conduct in the event they suffer a short, are grounded to the grounding wire also, by means of bonding with a grounding clip, or grounding screw, as previously mentioned.

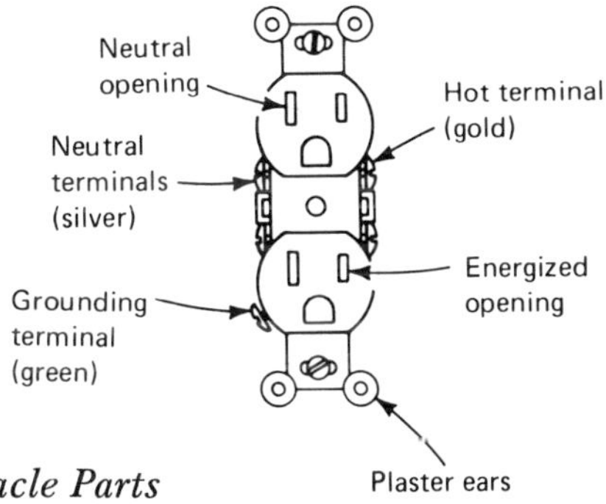

Figure 11: *Receptacle Parts*

*Wire Nut is the registered trademark of Ideal Industries, Inc.

Only grounding-type receptacles may be used in new work. The same holds true for old work, unless replacement of an existing non-grounding receptacle is all that is being done, *and* there is no grounding conductor in the outlet box and no convenient means for grounding. Then, replacement must be made with a non-grounding type receptacle. A ground should never be offered at any receptacle unless there is really a grounding conductor connected.

Receptacles are polarized. Therefore, the slotted openings for plug caps are of slightly different sizes, to prevent insertion of polarized plug caps in a reverse manner. Normally, as with lamp cord plug caps, little care is taken as to which of two ways the plug is inserted. Certain other appliances, however, are wired internally so that the on-off switch completely isolates internal components from power when the appliance is turned off. This can be accomplished only if the plug cap is designed so that it may be inserted in only one way. Manufacturers flange the neutral spade of polarized plug caps, making it larger in cross-sectional height than the energized spade. Thus, it can only be inserted into a polarized receptacle in one way: hot to hot and neutral to neutral. You may have noticed this on your TV or stereo system. Therefore, it is quite important that receptacles be correctly wired to preserve the integrity of polarized equipment. The "hot" slotted opening of a receptacle is the smaller one on the right-hand side of the receptacle as you look at it. This corresponds with the gold terminal screws being on the right and the white terminal screws being on the left. This assumes that the receptacle is installed with the opening for the grounding prod of the plug cap to the bottom. Installed upside down, of course, the receptacle has its left-right relationships reversed, but it is no less effective.

Figure 3, our cable routing diagram, indicated the use of 14/3 for wiring of the switched receptacle in the living room. Figure 12 illustrates how terminations would be made between the controlling switch and the receptacle under control using the 3-conductor cable. As the diagram shows, the third conductor, red in this case, represents the tie between the load side of the switch and the hot side of the receptacle. Black, our constant hot conductor, feeds on through the receptacle outlet box by being connected to the corresponding black conductor of the continuing cable. In the switch outlet box, the white neutral conductor is connected to feed through to the receptacle. The neutral is not terminated in the switch outlet box, as there is no load requiring the presence of a neutral. Additionally, we see the grounding conductor's continuity assured through the use of an approved connector. Two pigtails could have been formed, and one of the resulting pigtails could then be bonded to the box if the outlet box is made of metal.

Note in Figure 12 that both parts of the duplex receptacle are "on" or "off" at the same time. Sometimes, when it is desired to have half of the

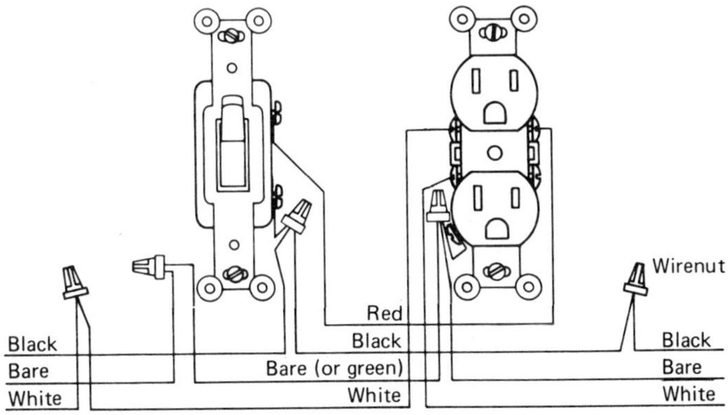

Figure 12: *Wiring of Switched Receptacle Using 3-Conductor Cable*

switched receptacle hot at all times, the receptacle is "split-wired," resulting in one half being controlled by the switch and the other half being energized at all times. This is easily accomplished by removing the factory-manufactured jumper between the two hot screw terminals and extending a pigtail from the connected joint of the two black wires on to the remaining hot screw terminal. This can be especially convenient when a lamp and a clock are on the same table and it is desired to switch off the lamp upon leaving the room without de-energizing the clock.

In considering terminations pertaining to receptacles, it is appropriate to examine wiring connections relating to multi-wire circuits and a special precaution that is associated with their use.

First, let's call upon Article 100 of the Code for a definition of a multi-wire branch circuit. It tells us that it is one "consisting of two or more ungrounded conductors having a potential difference between them, and a grounded conductor, having equal potential difference between it and each ungrounded conductor of the circuit, and which is connected to the neutral conductor of the system."* Had we run 3-conductor cable to *all* of our kitchen receptacles, we could have set up two individual circuits, as required by the Code for small-appliance kitchen receptacles, but yet have run only one cable to all receptacles.

Quite often, multi-wire circuits can be used to economical advantage for other household circuits as well. Prevalent in commercial work, they can also find use in residential work when the area to be served is some distance from the panelboard and it is desired to run one cable, instead of two, to serve two circuits.

---

*Reprinted by permission from NFPA 70-1981, National Electric Code®, copyright © 1980, National Fire Protection Association, Quincy, Mass.

Before going on, perhaps some comment is in order concerning the use of the word 'identified' in the Code discussion of an identified grounded conductor. 'Identified' means that the grounded current-carrying conductor referred to is covered with white or grey insulation, the mark of its being grounded, as explained in Article 200 of the Code. By regulation, the colors white and grey are reserved for those wires that constitute the return path to ground. However, an exception to this is when a cable is used as a switch leg from a fixture outlet box that is "fed hot." Then white or grey is the color of the wire that feeds the switch hot, and black is the color of the switched leg back to the fixture. Article 200 spells this out for us in a paragraph covering "Use of White or Natural Grey Color."*

A multi-wire circuit offers more than one hot wire. In a commercial application that utilizes 120/208-V 3ϕ 4-wire branch circuits, three hot conductors, all out of phase with each other, are extended out into the building from the branch circuit panelboard, along with a common neutral. From any one of the hot wires to neutral, there is a potential difference of 120 volts for lighting and receptacles. Between any two of the hot conductors, however, there is a potential difference of 208 volts, owing to the phase relationship between the two. Thus, this circuit offers three sources of 120 volts and actually constitutes three circuits, each 120 volts, but derived from different phases. If all three circuits were derived from the same phase, the neutral would be overloaded, but the out-of-phase relationship causes one to offset the other at the neutral point, and the neutral merely grounds the imbalance remaining. And this insures us that the neutral will never be loaded any more heavily than the most heavy loaded solitary hot conductor. The economies are obvious: two wires are saved, in contradistinction to three neutrals and three hots in conventional two-wire circuits. Loads are balance-connected as the multi-wire circuit makes its way through the building, so that more or less the same calculated load is impressed on each circuit.

Residential wiring is single-phase, not three-phase, as in commercial buildings, so two hot wires are brought out, instead of three, along with a common neutral. Three-conductor cable with ground is employed for residential multi-wire circuits, and the insulation colors are black, red, and white, with a bare or green grounding conductor. White, the identified conductor, is, of course, grounded, and black and red are the insulation colors of the two hot wires. There is a potential difference of 115 volts between neutral and either black or red, but a difference of 230 volts between black and

------ * Reprinted by permission from NFPA-70-1981, National Electric Code®, copyright ©1980, National Fire Protection Agency, Quincy, MA.

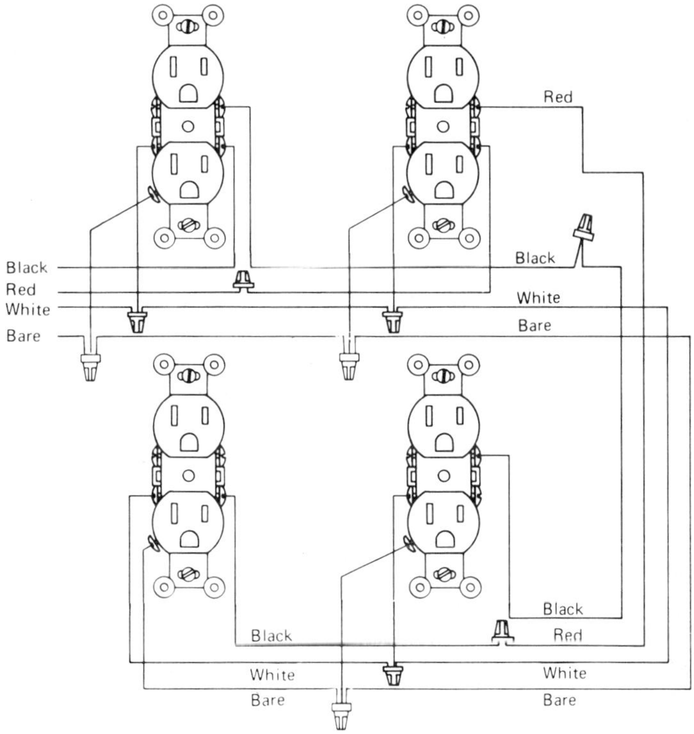

Figure 13: *Receptacles on 115/230-volt 1∅ Multi-wire Circuit*

red. We might decide to use this 3-wire circuit and alternately connect first one receptacle to black and then the other to red as we make our way around the kitchen or room, so that a different circuit appears on every other receptacle.

We will now consider the terminations and a special precaution associated with receptacles on a multi-wire circuit. Because a common neutral is used, care must be taken that the neutral is terminated at each receptacle in such a manner as to insure that removal and replacement of a receptacle at a later date will not result in isolation of "downstream" receptacles by opening up the neutral while loads are connected on the other half of the multi-wire circuit. Touch and go terminations, commonly used in ordinary two-wire circuits, cannot be employed for the neutral in multi-wire circuits, as removal of a receptacle would necessitate removal of the two ends of the neutral conductor.

Figure 13 shows a multi-wire circuit with receptacles. Note that leads to the neutral terminal on each receptacle are actually pigtails created by the formation of a joint in the outlet box and the addition of a 6-inch piece of white insulated wire to reach out to the neutral receptacle terminal. With such an arrangement, removal of the receptacle later on will not open the neutral of the common neutral circuit. The same holds true for the grounding wire, where a Wire Nut®* is used, with a pigtail extended to the grounding terminal.

The practical aspects of the necessity for jointing the neutral at each receptacle can be better appreciated when we look into the future and see someone attempting to replace a worn or damaged receptacle on a multi-wire circuit. Two breakers in the panelboard are labeled "kitchen circuit." Mr. homeowner turns off the one that puts out a lamp plugged into the receptacle he intends to replace. The receptacle still works, but it is worn, and connection at the plug cap is poor. Although he has de-energized the circuit that the receptacle is on, the other half of the multi-wire circuit, protected by *another* breaker, is still in operation, and the common neutral is conducting current. With the neutral jointed, he can remove the white wire terminated on the receptacle without disrupting operation on the other circuit, and without shock hazard. Had touch and go been employed instead, removal of the first white wire would have produced an arc, an outage on the other circuit, and possibly a shock to our unsuspecting friend.

Your municipality probably requires simultaneous actuation of both breakers on a multi-wire circuit in the event of reaction to overcurrent on either circuit or manual shutdown. This is good practice, as it assures that all energized wires entering an outlet where multi-wire circuiting is employed will be off when either circuit is disconnected. Two-pole breakers or breakers utilizing handle ties may be used to meet this requirement, and two-pole pullouts in fusible panels can afford the same protection.

TERMINATING CONDUCTORS AT SWITCHES

A single light switch employed for the control of an overhead fixture or wall receptacle is wired into the circuit so as to interrupt power to the fixture or receptacle outlet. This can be approached with power roughed into the switch outlet box, or power roughed into the fixture outlet box. One cable into a switch outlet box indicates an overhead feed to the fixture. The white wire to our switch is connected hot in the fixture outlet box, and the black wire is switched and wired back to the black wire on the fixture or re-

*Wire Nut is the registered trademark of Ideal Industries, Inc.

ceptacle terminal. As noted earlier in the discussion of wiring for receptacles, Article 200 permits the use of the white wire as a feed for switches when the hot wire appears in the outlet box with a controlled fixture or receptacle.

If our feed is roughed into the switch outlet box, then, of course, the switch breaks the two black wires, and the whites are jointed together with a Wire Nut®*.

Multiple switches at one location makes for an interesting observation concerning wiring, namely, that each of the switches is usually fed from the hot circuit entering the outlet box that contains the switches. Power jumpers from one switch to the other frequently are the means by which each switch is fed. Each switched-leg black wire is connected loadside of its respective switch, and all whites are jointed together for common gathering of neutrals. Figure 14 illustrates how our 3-gang switch arrangement of Figure 3 might be terminated.

Switch #1 controls the outside fixture by the front door, Switch #2 controls the lamp post at the end of the walk, and Switch #3 controls our

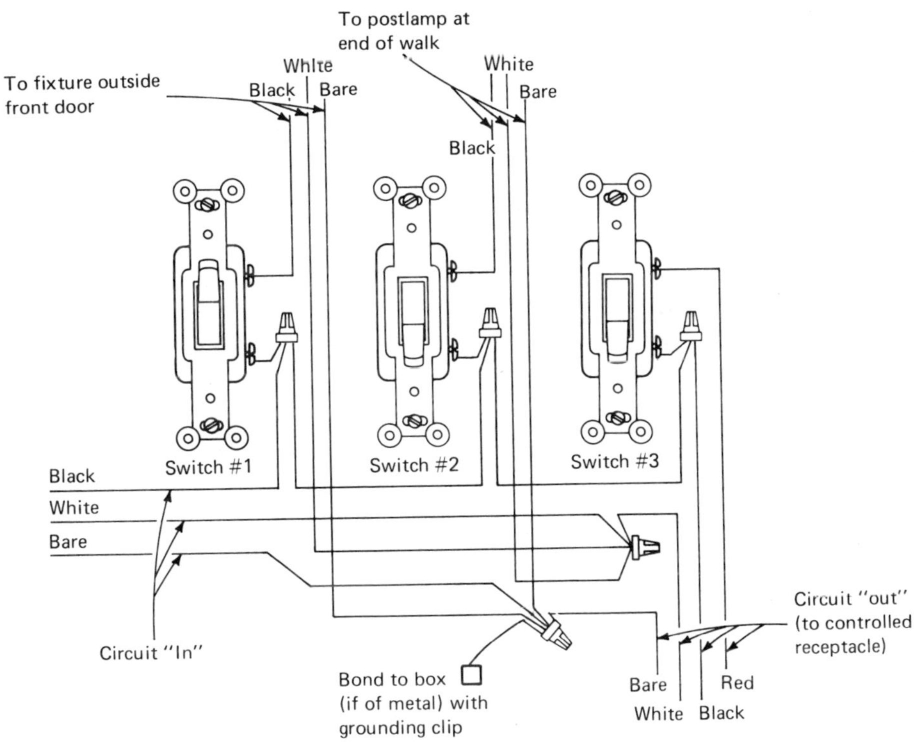

Figure 14: *Wiring of Multi-gang Switches*

*Wire Nut is the registered trademark of Ideal Industries, Inc.

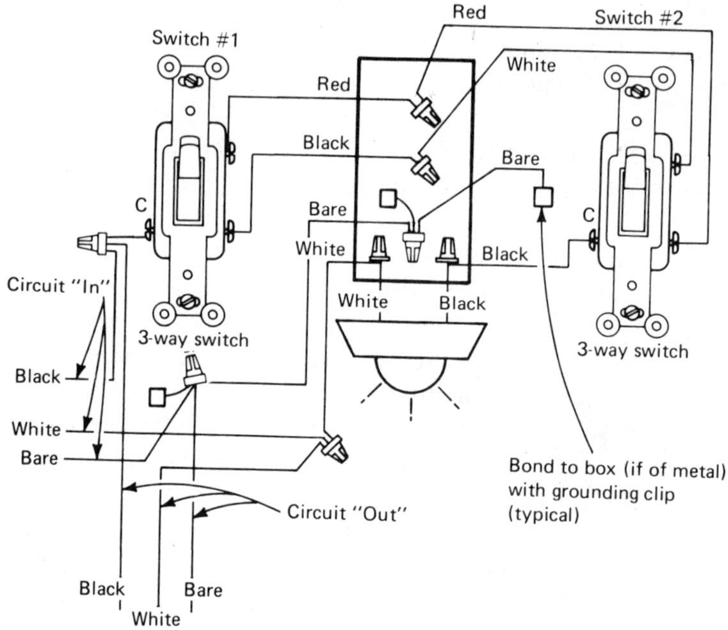

Figure 15: *Wiring of 3-way Switches with Feed to Switch Outlet*

switched receptacle by the front door. (See Figure 3.) Four wires, three insulated and one bare, constitute our circuit continuation cable, with the red insulation representing the switched leg to our controlled receptacle.

It is often desirable to control room lighting from more than one location. Rooms with multiple entries may have switches at each door into the room to permit lighting the room before entry. Where two switches are used to control the same outlet, 3-way switches are necessary. When it is desirable to utilize three or more switches for control of the same outlet, two 3-way, and the balance in 4-way, switches are required. Figure 15 shows typical connections for an overhead fixture, perhaps in a dining area or den of a home, controlled from two locations through the use of two 3-way switches.

Proper interconnection of our colored wires in the 3-wire cables between switches is important for preserving the integrity of the black wire connected to the fixture. Article 200, while permitting us to use white as the feed to a switch, also requires that black be the energized conductor back to the fixture when the hot feeds through the fixture outlet to Switch # 2 outlet, as it does in this case. White and red are used to feed Switch #2 from the fixture outlet box on two terminals that are alternately selected by throwing the toggle and connecting first one and then the other with the common terminal, marked C. Black then conducts power back to the black fixture lead wire. Switch #1 has its common terminal, marked C, fed hot, and alternately places power on first one, and then the other, of the remaining terminals, as the toggle is moved up and down. Thus, the light can be energized from either location, regardless of how the switch was left at the opposite location. From a roughing standpoint, three cables would be entering the outlet box at Switch #1: hot in, hot out, and the 3-wire switch leg, wired through the fixture outlet box onto Switch #2. Two cables enter the fixture outlet box: one 3-wire cable from each switch. And one cable enters the outlet box for Switch #2: the 3-wire cable from the fixture outlet box.

It might have been more convenient to route our hot circuit feed into the fixture outlet box, and were this the case, our connections for these 3-way switches would be as shown in Figure 16.

In the figure, 2-conductor with ground cable is used as the dropped feed to Switch #1, and white feeds black hot through the outlet box for Switch #1 to the common terminal of Switch #2. A 3-conductor cable interconnects the two 3-way switches. The wires tying the switched terminals of the two 3-way switches together are known as *travelers*, because the circuit back to the lamp in the fixture travels over one or the other of them

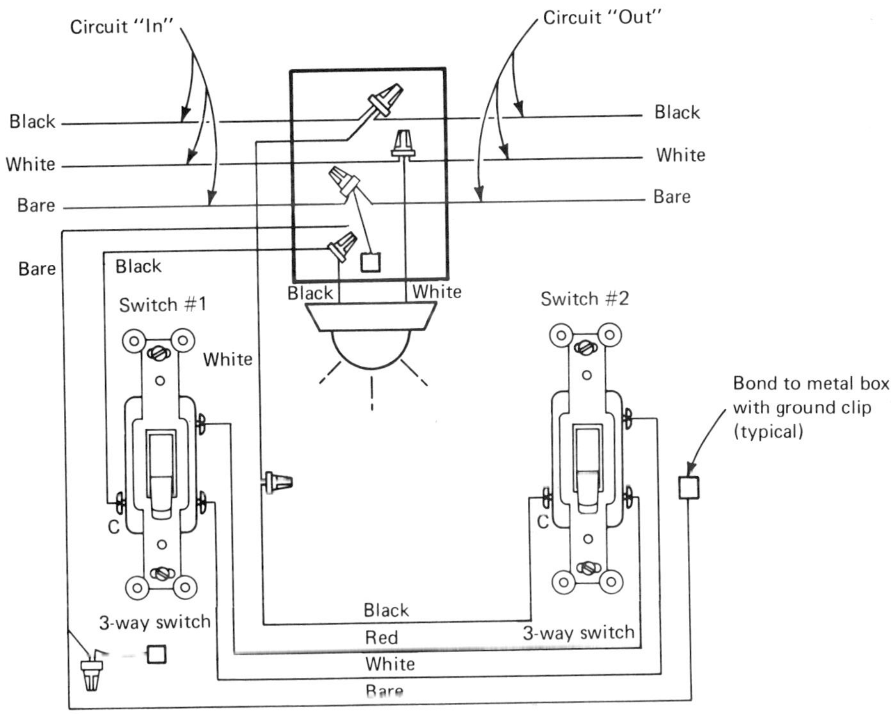

Figure 16: *Wiring of 3-way Switches with Feed to Fixture Outlet*

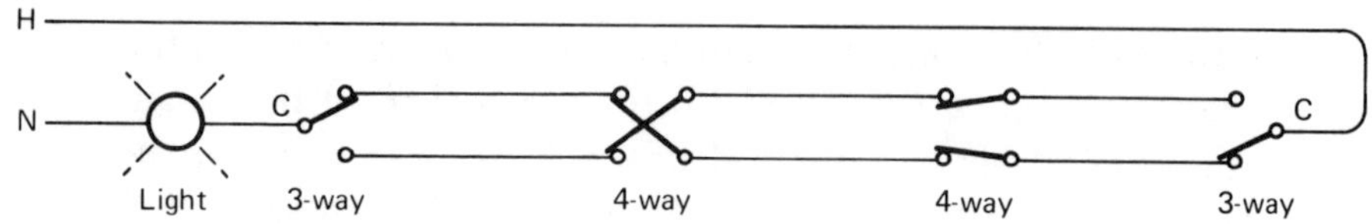

Figure 17: *4-way Switches in Light Circuits*

(in this case, red or white), depending on the position of the toggles at each switch. The integrity of the black wire is preserved in the outlet box for the fixture, and the black wire is connected to the fixture black lead. White lead off the fixture is connected into the white neutral.

The necessity to control an outlet from three or more locations imposes a requirement for the use of 4-way switches. Two 3-way switches is the maximum number permissible for the control of any one outlet from multiple locations, with the balance of switch requirements being met by 4-way switches. Our two 3-ways would be located on the outer end of the switch control loop with however many 4-ways we needed in between.

Figure 17 is a schematic wiring diagram of the 3- and 4-way switches in a typical light control circuit in a residence. The 4-way switch cross-connects travelers in one position and straight-connects them in the other. Four terminals appear on the switch, and generally, two of them are darker than the other two, indicating touchdown points for the travelers in one cable, with the other travelers in the remaining cable going on the lighter terminals. As is evident, the sole function of the four switches in this lighting control loop is to permit the selection of alternate paths from any one of the switch locations that will open or close a circuit back to the lamp in our fixture under control. As shown, the schematic diagram has the lamp lit. Operating any one of the switches will extinguish it, and the subsequent operation of any one of the switches again will relight it, and so on. For this reason, toggles on 3- and 4-way switches are not engraved with "on" or "off," as either position could be "on," depending on toggle positions in other switches in the loop.

Additional fixtures in the same area could be wired in parallel with our one fixture shown in the schematic, for simultaneous control of multiple fixtures from multiple locations. Should a dimmer be desired, it must be of the 3-wire type and installed in place of one of the 3-way switches. Only one dimmer can be used, with the rest of the locations utilizing switches only. Usually, the common wire of a dimmer is distinguished from the other two wires and would be connected as are our common switch terminals in the schematic. Let us now take a look at a diagram that fleshes our schematic into actual switches. In Figure 18 we feed Switch #1 with white dropped from the fixture outlet box, and black represents our switched leg back to the lamp in the fixture. Notice that black comes from Switch #4 and is jointed to other black insulated conductors in other sections of cable, as it feeds through switch outlet boxes, to preserve the integrity of "black back

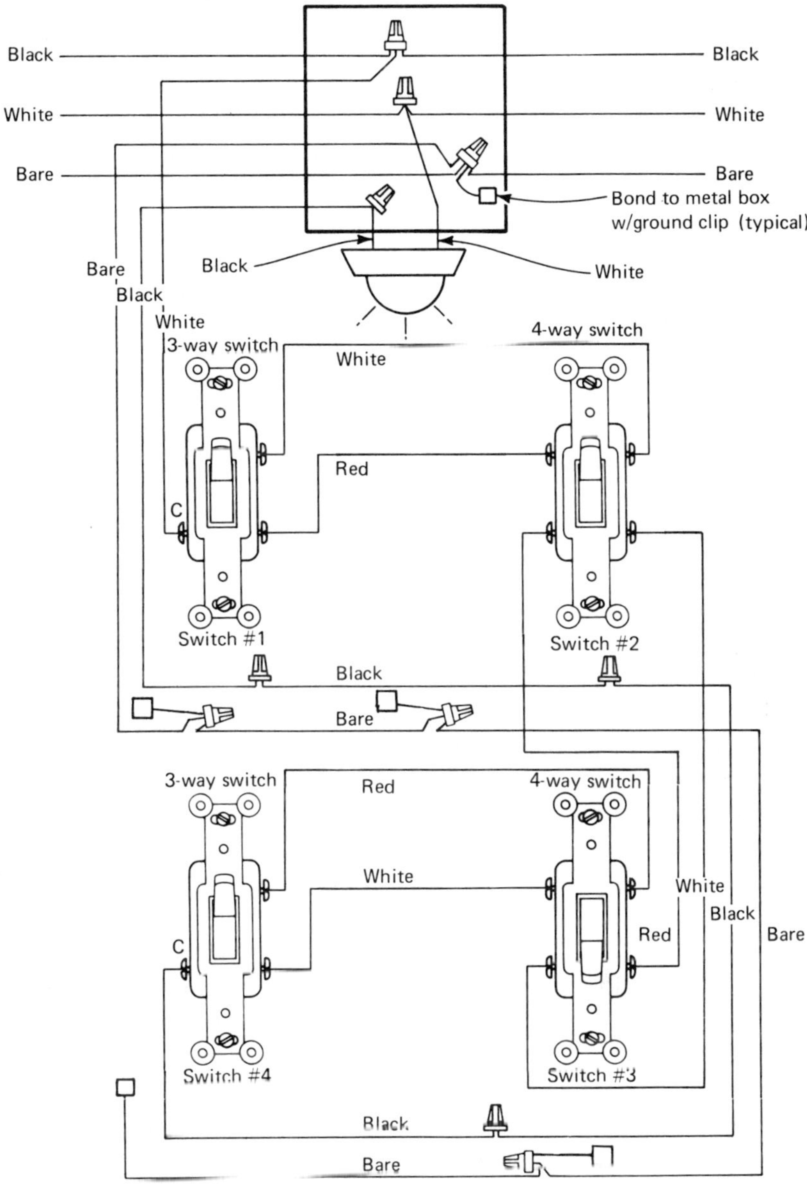

Figure 18: *Wiring of 4-way Switches with Feed to Fixture Outlet*

to the fixture." Therefore, white and red become our travelers when Switch #1 is fed first. We could have fed Switch #4 first by jointing white through the outlet boxes to the common of Switch #4; then red and black would have become travelers, and "black back" from the common of Switch #1 would have lit our lamp in the fixture. The key wire is the black one, because we should have a black wire entering the fixture outlet box to mate with the black lead wire from the fixture, so its white lead can be appropriately terminated with the system neutrals.

Exceptional versatility is afforded through the use of 3- and 4-way switching, permitting control of lighting in an area from several locations, so as to insure that entering a darkened room is unnecessary, which would not be so if we could not turn on the light at the doorway before entering. Of course, our controlled outlet could have been a receptacle outlet into which a lamp is plugged, as permitted by the Code in all habitable rooms except the bath and kitchen.

REMOTE-CONTROLLED OUTLETS

During remodeling or new construction, it may be desirable to increase the scope of the electrical work to include the ability to control selected outlets from one or more specified locations. An ideal time to get the necessary wiring into the walls to facilitate selective control of certain outlets for convenience or decoration is when walls are open, or when significant patching and painting is to be performed anyway. Consider floodlighting around the house, with a switch in the utility room, by the back or front door, and perhaps even in the master bedroom to serve as a convenient point for energizing floodlights in the night if a disturbance is noticed. Recalling from the preceding pages the necessity for 3-conductor cable between multiple switch locations and the use of 3- and 4-way switches, we observe that during remodeling or renovation, necessary cuts in the walls could be made in order to get by studs and joists with a minimum of additional patchwork. Each 150-watt bulb represents about $1\frac{1}{3}$ amperes on a 115-volt circuit, so a special circuit should be required if total floodlight power exceeds 50% of an available circuit that is being considered for tie-in.

Split-wired receptacles might be a desirable feature in the master bedroom, too, for turning off the television across the room from beside the bed. Here 3-conductor cable would be used, with the third conductor being controlled by a switch by the bed and connected to just one half of each receptacle for remote control.

Another ideal application of the split-wired receptacle has to do with Christmas candles placed in the windows during the Christmas season. Use of extension cords for these candles can be eliminated by the installation of a receptacle under each window where a candle would be placed. One half of each special Christmas candle receptacle would then be connected to one switch, permitting all electric Christmas candles to be turned on and off at one time from one location. Extension cords are eliminated, and candle cords can be shortened to 30 inches or so, just long enough to reach from the windowsill to the receptacle.

An independent circuit should be used and routed first to the switch location. From there 3-conductor cable would be used so that half of the receptacles could be hot at all times.

If our electric candle receptacles were located in a dining room, pantry, or kitchen, they would be in addition to those used to meet the requirements for heavier receptacle circuiting in these rooms. These receptacles should be in addition to the ones required to be circuited under Article 220 of the Code. Such a precaution would insure that the integrity of our special circuiting and the integrity of receptacle placement requirements for these higher activity rooms were preserved.

Other desirable applications for remote-controlled outlets no doubt come to mind. Much convenience and flexibility in the use of receptacles can be obtained with very little difficulty and extra cost if electrical plans are made to coincide with remodeling, new work, or other renovation activities.

CONNECTING COMBINATION BATH FANS

A very nice feature of some fans offered for use in bathrooms is that they can combine exhaust with light and heat into one unit. An electric resistance heating element, frequently in the range of 1,200 watts, is incorporated into the ceiling unit. A special-purpose 20-ampere 115-volt circuit must be extended separately from the panelboard with #12 wiring to the outlet box for the switches that will be controlling the ceiling unit. The fan manufacturer provides a double-gang switch assembly with his fan, oftentimes consisting of three switches and one pilot light. A separate switch is provided for fan, heat, and light, and the pilot light indicates when the heater within the fan unit is energized.

Four conductors plus a bare grounding conductor (or green) are required from the switch outlet box to the ceiling unit. When all components of the unit are in operation, the bathroom is illuminated from the ceiling by

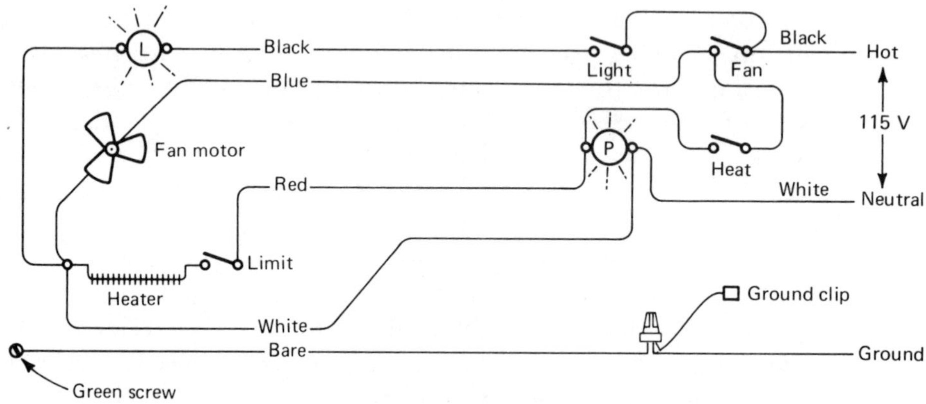

Figure 19: *Wiring Schematic for Light/Vent/Heat Unit*

virtue of an incandescent bulb fixture within the unit. Space exhaust occurs through the fan part of the unit, and heating takes place as a small portion of the exhausted air is directed back into the bath over the hot resistance element. Any one of these individual functions may be indexed for operation independently, except that in most units heating is interlocked so that it cannot occur unless the fan is running. This insures airflow over the heating element. The interlock is generally accomplished at the switch, as we see in Figure 19. Of course, the heating unit does incorporate a thermal high-limit device, designed to de-energize the heating element if excessively high temperatures occur, but the interlocking of the fan prevents the necessity for its coming into play. Figure 19 depicts a schematic of our combination ceiling unit circuit. Switches and pilot light are on the right, with the unit at the left. Closing the fan switch completes the power circuit to our fan motor and places our heat in the "ready" position. Closing the heat switch completes the power circuit to our heater in the unit, subject to fan operation, and energizes the pilot light. And, of course, closing the light switch causes our fixture to light. Should the fan switch be opened with the heat on, the heater will stop, to prevent high temperatures that could be damaging to the heating element from occurring as a result of cessation of airflow.

On the wall, our combination bath fan unit control center might appear something like Figure 20. Switches could be placed for either horizontal or vertical throw. The lens in our pilot lamp might be red or green, and its purpose is to draw our attention to the fact that the heater portion of our ceiling unit is indexed to operate. Generally, a double-gang switch box is required to be installed during rough-in to accommodate the manufacturer's switch package.

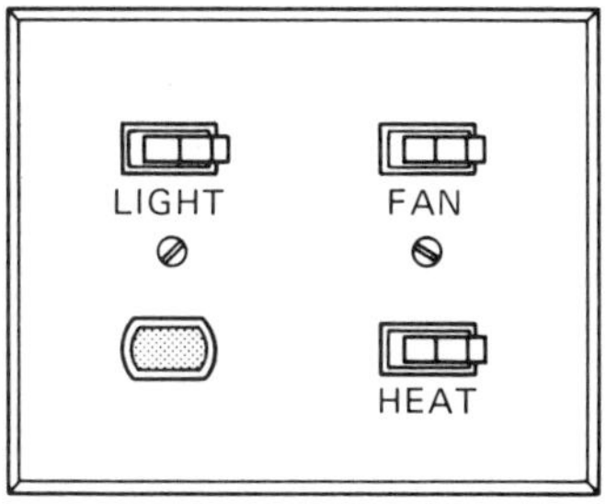

Figure 20: *Light/Vent/Heat Control Switches*

Oftentimes, the switches and light are pigtailed with stranded wire to facilitate Wire Nut®* connections; or, separate screw terminals may be the means for terminating our wiring, in which case special care is to be exercised to insure cross-connecting in accordance with manufacturer's diagrams, to afford proper interlocking and correct switch-to-function identity. Dual-fan units, with a separate fan for heating, may be directly indexed for heating, since the heating fan is energized from the same switched leg. Many of the newer units use two fans, one for exhaust and a separate one for heating.

WIRING AT FIXTURES

Fixture hanging and connecting is a part of the electrical installation that can contribute significantly to the finished appearance of a room. Often, owners of custom-built homes reserve the right to select their own fixtures, working toward a wholesale cost allowance which the electrical contractor may have quoted into the job. This is as it should be, because fixture preference is a personal thing. People's tastes differ, and their interior decorating plans weigh heavily in decisions relating to fixture selection.

Mounting hardware is packed with the fixture and will usually include hanging bracket, fixture screws, and Wire Nuts®*. Fixture wiring leads will have one wire, the neutral conductor with white insulation, and one, the energized conductor with another color, usually black. Fixtures with more than one bulb probably will have one black and one white wire for each bulb socket. All leads of the same color will be connected together with the same color wire roughed into the fixture outlet box, so all bulbs will light simultaneously. It makes a difference which wire of the fixture is connected to

*Wire Nut is the registered trademark of Ideal Industries, Inc.

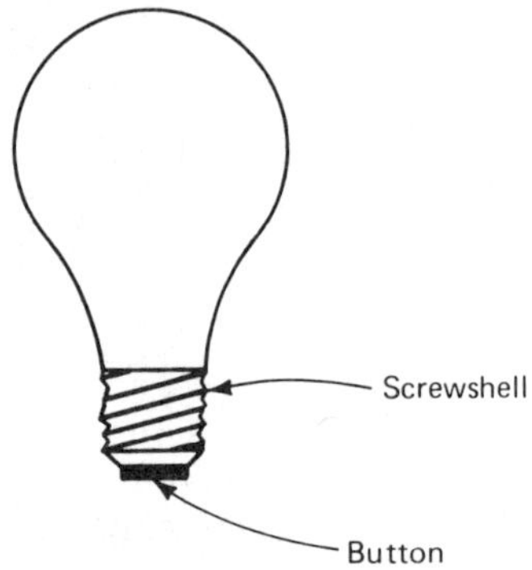

Figure 21: *Contact Points of an Incandescent Bulb*

our switched hot leg from our wall switch. Current flow through the incandescent light bulb is from the button-type contact point at the very base of the bulb, through the filament inside, and back to the threaded metal screw shell base. From an operational standpoint, the bulb will work regardless of which fixture lead is connected neutral and which is energized. But from a safety standpoint, connections do make a difference. We draw as our reference Code Article 410, which, under "polarization of fixtures,"* points out that the identified grounded conductor shall always be connected to the screwshell. In permanently connected light fixtures, the white lead provides for this, as it is connected internally to the screwshell side of the bulb socket. The reason for this requirement is so that the energized part of the bulb socket is (1) kept to a minimum insofar as energized area is concerned, (2) prevented as much as possible from being accidentally touched during bulb changing, and (3) isolated to the greatest extent from contact with the exposed metal portion of the fixture housing.

Bare or green grounding leads of a fixture must be connected to the grounding conductor of our circuit in the fixture outlet box. This insures that the exposed metal parts of our fixture are kept permanently grounded. Consequently, stray current cannot appear on any parts of the fixture that are not supposed to be energized, and the operation of our circuit overcurrent protective device, breaker or fuse, is facilitated in the event voltage is accidentally introduced onto the surface of the fixture through an electrical fault.

Porcelain bulb holders find widespread use in areas of the house that do not require anything other than a purely utilitarian light fixture. Such areas

*Reprinted by permission from NFPA 70-1981, National Electrical Code®, copyright © 1980, National Fire Protection Association, Quincy, Mass.

might be basements, attics, garages, utility rooms, or crawl spaces. A pull-chain type might be used, or a model without a pull-chain might be preferred and switched from the wall. Such fixtures generally feature screw terminals rather than wire leads, but one will be the grounded terminal, essentially white in color, while the other will be dark or gold in color, representing the termination point for the energized conductor.

Figure 22 shows our branch circuit feeding through the fixture outlet box and our switch cable connected in so that its black wire represents the return to the fixture from the switch. Field-installed pigtails or leads make connections to the fixture much easier. It can be very difficult to attempt to terminate multiple wires on the screw terminal of a fixture like this. Even if accomplished, mounting of the fixture and subsequent forcing of the conductors back into the fixture outlet box imparts strain on the screw terminal and frequently results in damage to the interior of the fixture's electrical parts. It is far better to install pigtails of the same size and material as the circuit conductors and terminate just one wire on each terminal. Then the balance of the wiring can be tucked into the outlet box, and mounting of the fixture necessitates flexing of the pigtails only.

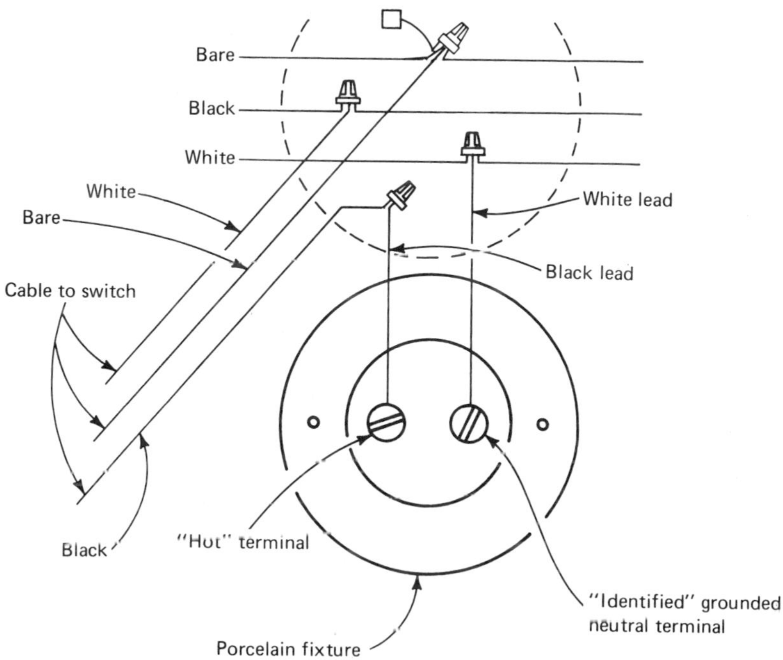

Figure 22: *Wiring of Remotely Switched Porcelain Fixture*

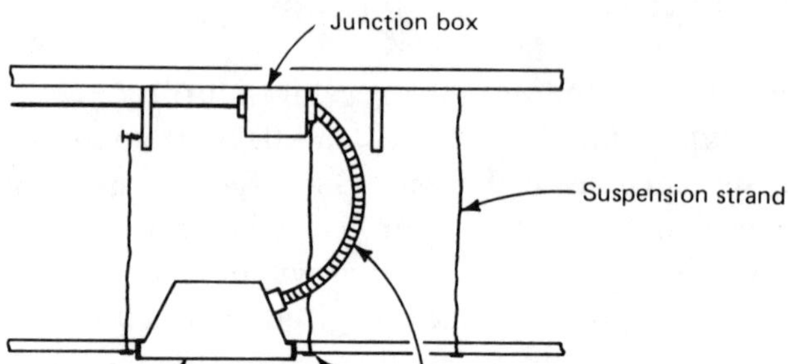

Figure 23: *Wiring to Fluorescent Fixture in Suspended Ceiling System*

Fluorescent fixtures find extensive use in today's homes also, particularly in the kitchen, recreation room, and shop areas. They can be surface-mounted or flush-mounted, and they afford much light for the wattage consumed. A word of caution needs to be given concerning wiring that enters the interior of the fluorescent. Ballasts are used with fluorescents to strike the arc and regulate current flow, and in normal operation, they generate heat within the fixture casing. Generally, wiring within the fixture itself must be rated at 90°C, to insure that the insulation on the conductors will not break down in the higher temperatures that are found inside fluorescent fixtures.

Such temperature ratings are higher than those of the conductors within our cable used for circuit wiring, so conversion is necessary before entering the housing of the ballast. In exposed work, or above lift-out, accessible ceilings, our cable terminates in a junction box, and flexible metal conduit with high-temperature conductors is extended into the fixture itself, using a flexible metal conduit-connector through a knockout of the housing. Figure 23 shows a drop-in fluorescent fixture installed in a suspended ceiling system. Inverted T-bar framing holds the ceiling tiles and supports our fixture, which occupies a space where a ceiling tile would otherwise have been inserted. Note our junction box in the joists above, with non-metallic-sheathed cable coming to it through the wooden members.

In open work, such as basement ceilings that are unfinished, a surface-mounting type fixture might be used, mounted against the joists. But we would still make use of the remote junction box and flexible metal conduit with high-temperature wiring in between. Article 402 of the Code lists fixture wires from which we may choose those that are suitable for the purpose.

CHANDELIER HANGING

Special attention should be focused on installation of chandeliers and the technique used to prevent unnecessary twisting of fixture wires during hanging. An all-thread nipple is used to support the chandelier and adjust the dimension of the space between the ceiling and the threaded collar onto which the escutcheon nut is installed. The fixture wiring to the chandelier must pass through the all-thread nipple and up into the open space of the clevis coupling if the fixture is to be hung from a fixture stud within the outlet box. If the fixture is to be hung from a box-mounted, center-tapped fixture strap, the wiring must pass through the nipple up into the outlet box. The all-thread nipple makes it unnecessary to rotate the fixture itself in order to achieve the threaded mounting; rotating the fixture usually imparts strain on the wiring, causing knotting and insulation damage, and sometimes conductor fatigue. Instead, the nipple is adjusted up or down to achieve the distance between the ceiling and threaded collar necessary to permit flush installation of the escutcheon cover. Then, the nipple is threaded down into the mounting collar. At the same time, the nipple moves downward, preserving the distance necessary to permit the neat, finished appearance of the properly installed ceiling escutcheon.

After rough-in and installation of ceiling material, the chandelier can be hung. As we approach the fixture outlet box, we note the circuit conductors within the box, ready to be terminated to our fixture. A four-foot stepladder is ideal for use during fixture hanging. Assuming a plain octagonal fixture box, our first step will be to install the cross-strap furnished with the hanging fixture. Then, the all-thread nipple can be threaded up into the prethreaded hole in the center of the strap.

Determine the height from the floor at which the bottom of the chandelier will be located. Usually, ample chain is provided with the fixture, and several links may have to be removed to have the fixture hang at the desired location. Hold the fixture in place if the owners are present, or if it is your home, have your spouse judge as you simulate the final hung height of the fixture by holding it below the outlet box. Remove the necessary number of links, and weave the fixture wires up through the hanging chain. Take care to straighten the links and hold them straight out, so that the wires have more slack in them than the chain does; we do not want any of the fixture weight to be imparted to the fixture wires. Next, back thread the escutcheon nut down over the collar, and follow that by passing the escutcheon down over the collar as well, and then let them both pass down the chain until they come to rest.

Get back on the ladder with the fixture in hand, and insert the fixture wires up through the threaded collar at the top of the chain and then through the all-thread nipple, already installed in the cross-strap. Bring the collar to the bottom of the nipple, and begin to thread the nipple down into the I.D. threads of the collar. After engaging a half inch or so of the nipple, bring the escutcheon up to the ceiling, and check it for fit. Does it fit snugly to the ceiling and permit sufficient thread to be engaged by the escutcheon nut to secure it up against the ceiling? If not, determine how much raising or lowering is required, and thread the nipple out of the collar. Then, with the fixture free of the nipple, turn the nipple enough to achieve the desired raising or lowering; bring the collar back to the nipple, and rethread. As you can see, all hanging and adjusting is done without twisting the fixture wires by merely turning the nipple free of the fixture up or down to accomplish adjustment and then turning the nipple into the collar of the fixture to accomplish mounting. Once the threaded collar at the top part of the fixture chain

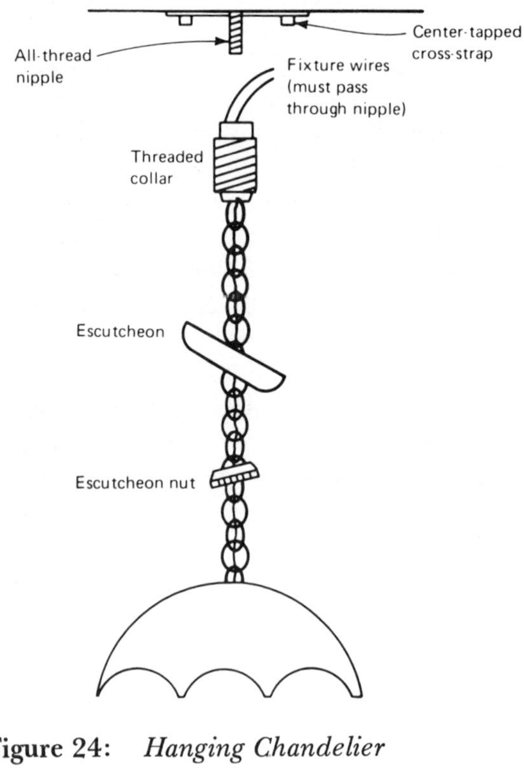

Figure 24: *Hanging Chandelier*

is firmly engaged, the fixture is hung, and joints can now be made between circuit conductors and fixture conductors, using, perhaps, the Wire Nuts®* that came with the fixture. Then, the escutcheon is raised to the ceiling, and the escutcheon nut is threaded onto the threaded collar to hold it firmly to the ceiling.

Figure 24 illustrates the various chandelier components. The fixture is shown ready to be hung, with the escutcheon and escutcheon nut resting halfway down the chain, ready to be raised into place after anchoring is accomplished between the nipple and collar. During hanging, be sure sufficient thread is engaged on both sides of the all-thread nipple to insure stability and firm anchoring. Needless to say, the switched leg to the fixture should be dead during installation.

GARBAGE DISPOSAL INSTALLATIONS

Many homes today feature disposals, and some attention needs to be focused on wiring methods employed in under-the-sink spaces where the disposal is to be electrically connected. Non-metallic-sheathed cable is not to be run exposed under sinks, because of the possibility of damage to the cable when items are moved in and out or when plumbing repair work is necessary. Earlier, we mentioned appliance stub-ups in connection with installation of cables. In most cases wiring connects to the back of the appliance and is concealed when the unit is in place. However, with disposals, wiring is exposed in the storage area under the sink. So conversion to a wiring method that will afford mechanical protection to our conductors is required.

Flexible metal conduit is well suited for use in this area. If a previously roughed-in wall stub-out has been brought through the back of the cabinets, a junction box may be installed over the cable where it enters the under-the-sink space, and the cable may be connected with a connector and the box mounted against the back board. Then metal flex with wiring installed can be connected between junction box and disposal wiring compartment. If our stub-out is not a switched leg, then a second section of metal flex can be extended to our switch outlet box and connected to control our disposal.

Figure 25(a) depicts a side view of a disposal with wiring installed. Metal flex is extended to the switch outlet if the location is to be on the front of the cabinet. An alternate location in the wall above would require just NM sheathed cable, as is used in general wiring throughout the house.

*Wire Nut is the registered trademark of Ideal Industries, Inc.

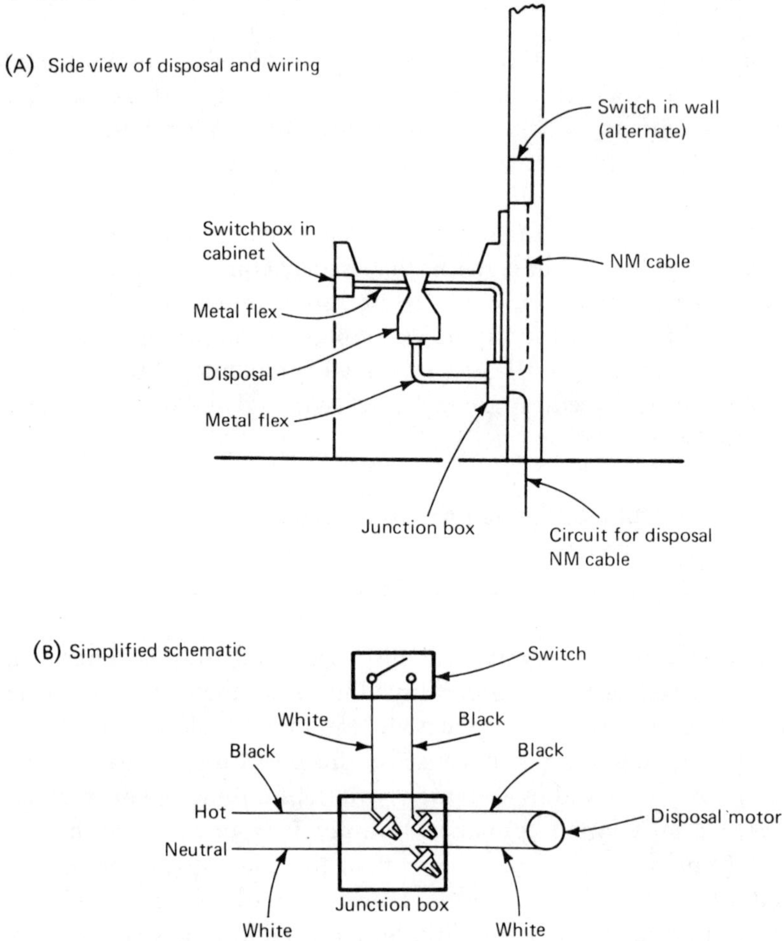

Figure 25: *Disposal Wiring*

Figure 25(b) shows the electrical connections for the circuit conductors. Notice that white is fed hot to the switch and black is reserved for the return to the motor leg.

The electrician's work on the disposal is performed after cabinets are in place and the plumber has mounted and piped the disposal unit. A trouble light is an essential tool, as under-sink spaces are dark and cramped. Care must be taken to insure that cables roughed in for disposal use are brought through the back of the sink counter when the counters are put in place. Otherwise, cables may be quite difficult to locate.

Occasionally, passage through the wall void for location of a wall switch to control the garbage disposal may be blocked by a horizontal brace. Where ceramic tile is used as the counter backsplash material, cutting the wall and notching the brace is precluded. At this point, installation of conduit on the outside presents itself as the likely alternative. From the basement or crawl space, penetration of the wall with a ¾-inch hole will allow access to the outside, where we can rise up to the wall switch box location in ½-inch EMT, using ½-inch SLBs on either end of the EMT to permit turning the corner. Cut another ¾-inch hole through from the outside, zeroing in on the center of the back of the switch box. Then, an EMT connector may be used to secure the conduit to the switch box.

The SLB is a conduit body similar to ordinary 90° conduit bodies, except that its design allows the EMT to hug the wall without the necessity of recessing the threaded neck of the conduit body. This is particularly helpful, as holes which penetrate the wall need only be large enough for the outside diameter of the conduit itself. A steel-threaded nipple can be screwed directly into the SLB and locknutted to the switch box through its rear knockout, for mechanical and electrical grounding integrity of the conduit system. Non-metallic cable extended to the wall penetration point in the basement or crawl space would enter a junction box, where its grounding conductor would be bonded to both the metal box and the grounding conductor installed in the raceway.

If space is limited above the ceramic tile, horizontal installation of the switch box for the disposal might be the best solution. Any of the "old work" type boxes can be employed where mounting within the wall cavity does not rely on fastening to a stud. Instead, open wall space is preferred, and box clamps secure the box from behind, causing the ears at either end of the box to pull down against the finished wall surface.

In installations that use conduit on the outside to reach the switch box, joints would be made in the crawl space or basement. The switched leg to the disposal would leave the junction box below in greenfield (flexible metal conduit) and rise up from below into the under-the-sink space. Access might be attained beside the cold water piping rising up to the sink. Connection to the disposal would be with a greenfield connector. Be sure to pull a green or bare grounding conductor in with the black and white wires. The grounding conductor must be terminated on the green grounding screw within the wiring compartment at the bottom of the disposal. Come up into the under-the-sink space close to a side partition or toward the rear with the piping. Then, flex over to the disposal in a 90° sweep. This will leave the floor area under the sink free for storage of miscellaneous cleaning agents normally found there. Length of greenfield should be supported every 4½ feet and within 12 inches of each box and fitting.

2

HEATING AND AIR CONDITIONING

HEAT PUMP WIRING

Let us now turn our attention to power and control wiring associated with our heating and air conditioning system. A few new twists will occur that represent somewhat of a departure from the straight cable pulls we've been making throughout the rest of the house. The use of conduit, both straight and flexible, appears on the scene as we begin to wire for the heat pumps.

Both the indoor and outdoor units require individual power circuits. These originated at our panelboard with runs of heavy cable, leading out through the framing members of the house. If the indoor unit is located in a utility room, we will have to drop our circuit from the ceiling joists overhead, set the disconnect, and terminate on the incoming lugs. As mentioned earlier, this disconnect may be a fusible switch or a fusible pullout. The switch opens the circuit by using a handle to operate contacts; the pullout offers a handle that permits removal of the pullout section, thereby extracting the fuse block and opening the circuit.

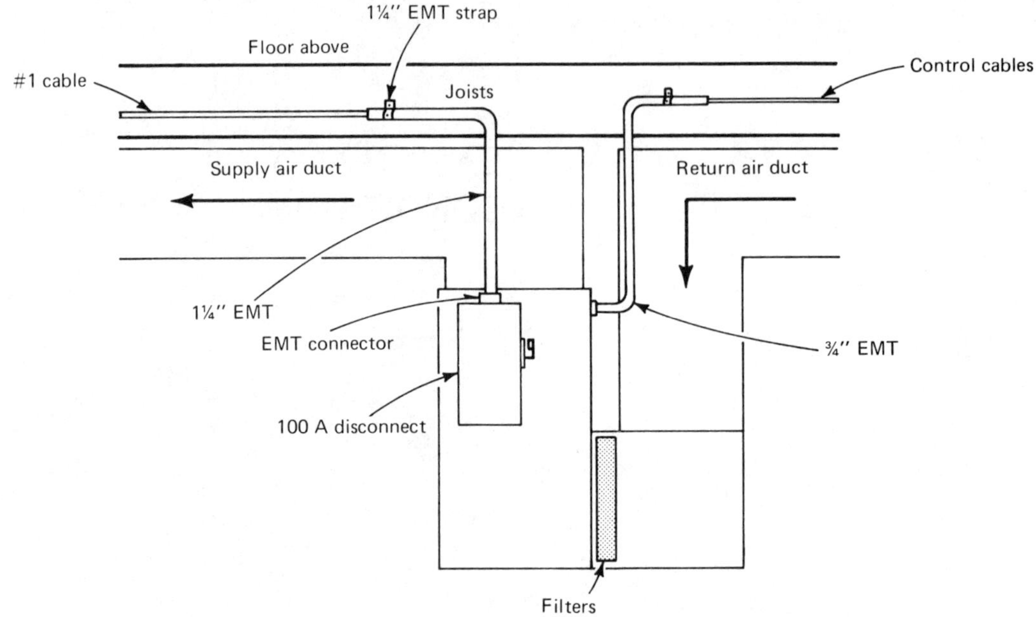

Figure 26: *Wiring to Indoor Section of Heat Pump*

Disconnects are available in standard sizes of 30 amperes, 60 amperes, 100 amperes, 200 amperes, and up. Fuses are manufactured so that they will not fit into a disconnect rated less than they are. The use of 110-ampere fuses would force us to choose a switch or pullout rated at 200 amperes, quite a bit larger and more expensive than one rated at 100 amperes. An exception in Article 220 permits listed overcurrent equipment to be applied up to 100% of its rating, and a wise decision here might be to go with the 100 ampere unit, fused at 100 amperes. This would establish branch circuit overcurrent protection at that level, and still permit our indoor unit to operate satisfactorily since actual current draw is less than 100 amperes. And this selection would be the practical choice when calculated conductor amperage requirements are so close to a plateau in manufacturers' standard disconnect size offerings. Although the conductors must be sized for circuit load plus 25%, we may protect them at any value *less* than that, so long as proper operation of the piece of equipment being served is not impeded in any way.

So we decide to use a fusible safety switch rated and fused at 100 amperes. The switch might be mounted on a wall within 6 feet of the unit, with flexible conduit bridging between them. Or we could mount our switch on the unit itself, so long as it is readily accessible and does not block access panels. Either way, we must drop from the ceiling joists in straight conduit and use an approved connector where we enter the disconnect enclosure. Figure 26 shows our disconnect positioned on the indoor unit itself, and our drop to the unit is with electrical metallic tubing (EMT), somewhat lighter and easier to bend than rigid steel conduit. Electrical metallic tubing, when written, is frequently abbreviated using the letters EMT, or the letters TW which stand for *thin-wall.*

EMT is available in standard 10-foot lengths and can be easily bent with a bending tool. The tubing mechanically protects the electrical cable as it drops down to the unit, where it might be susceptible to physical damage. It acts as a raceway for our power conductors and their jacket. Our conductors leave the loadside of the disconnect through the back, using a conduit nipple and bushing.

Control wiring would probably consist of two #18 gauge 8-conductor cables, one run to the thermostat in the occupied space and the other extended to the outdoor unit. Heat transfer with the outside air is accomplished by the weatherproof coil, fan and compressor section of the latter, which is interlocked to run with our indoor unit through the control cable. Usually, low-voltage relays, activated by our control wiring, energize the units in response to space temperature demands.

Now consider Figure 27, which illustrates wiring for the outdoor section of our heat pump system. Our decision was to use 50-ampere fuses, in accordance with the manufacturer's nameplate information, which gave 50 amperes as the maximum permissible fuse size. We select, therefore, a 60-ampere weatherproof disconnect, this time of the pullout type, featuring construction that prevents the interference of weather with its successful operation. Flexible, weatherproof conduit, or more properly, liquid-tight flexible conduit, is used outside between disconnect and unit, with special

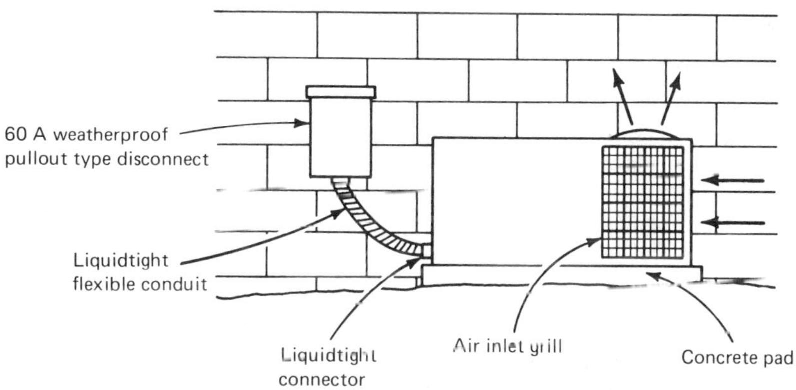

Figure 27: *Wiring to Outdoor Section of Heat Pump*

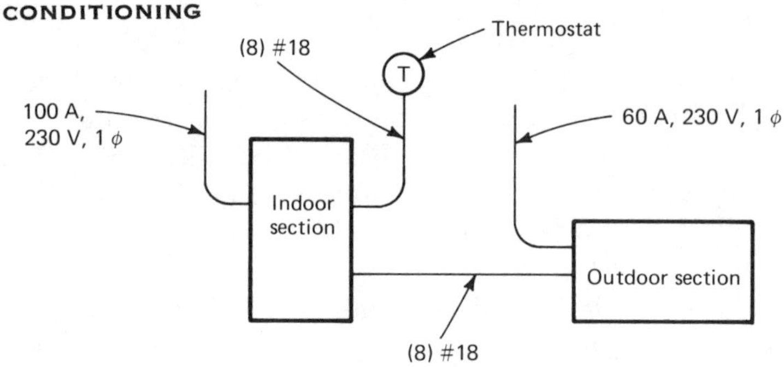

Figure 28: *One-Line Power and Control Wiring Diagram of Heat Pump*

connectors. Entry into our disconnect from the crawl space under the house is through the back of the enclosure, using a cable connector with cable extended through the wall. First we terminate conductors on the incoming lugs of the disconnect; then we extend our #6 power wires and control cable through the 1¼-inch liquid-tight into the unit, and terminate them on the proper lugs, in order to serve the fan and compressor located in this section. The disconnect may be anchored to the masonry wall with any suitable fastening devices. If we have a wooden wall, appropriate wood screws will do the same job.

Figure 28 is a one-line wiring diagram of our heat pump sections, interconnections, and power wiring. Both of our power circuits to the two pieces of equipment utilize large cables that are exempt from the requirement that they pass through joists. Article 336 of the Code permits cables of (2) #6 or (3) #8 and larger to be secured directly to the lower edge of joists. So if local inspections officials in your area assent to it, installation of such cables in crawl space or basements can be made without boring holes in joists.

HEATING PLANT

Some discussion is necessary regarding wiring for oil- and gas-fired heating systems. The hypothetical house we have been discussing utilizes a heat pump for year-round comfort control. But heat pumps are by no means the only heating equipment being installed today. Central air conditioning is not always incorporated into new house design, so consideration of wiring techniques for conventional heating systems of long standing use is appropriate.

The load center for a home using a fossil-fuel-fired furnace might be smaller than for one using a heat pump with supplemental electrical resistance heaters. Less electricity would be consumed in heating the home in very cold weather, but, of course, the expenditure for electricity would be replaced by an expenditure for oil or gas and could be just about the same annually, depending on the home and whether or not insulation was used effectively.

Heating plants can be located in basements, under the house in crawl space, in utility rooms, or in attics. Wherever they are located, a receptacle should be provided for the convenience of repairmen, and a light fixture must be provided. For wiring around the unit, suitable high-temperature insulated wire, rated at no less than 90°C, should be employed.

Whether of the hot water or warm air variety, the boiler or furnace should be on an independent electric circuit. In most cases, 20 amperes is sufficient. An in-sight, serviceman's disconnect will be located on or close to the furnace, and flexible metal conduit will be used to wire from one control to the other and to motors for burners, circulators, or fans.

If the heating plant is located in the basement, our overhead wiring to the unit will be non-metallic cable, and as we drop down to the disconnect, EMT can be used to best advantage. Figure 29 depicts an oil-fired warm air furnace, with the various controls connected with flexible metal conduit. The stack switch senses stack temperature to confirm whether the flame is ignited and to permit continuation of burner operation. It stops the burner motor in the event the flame is extinguished. The combination fan and limit

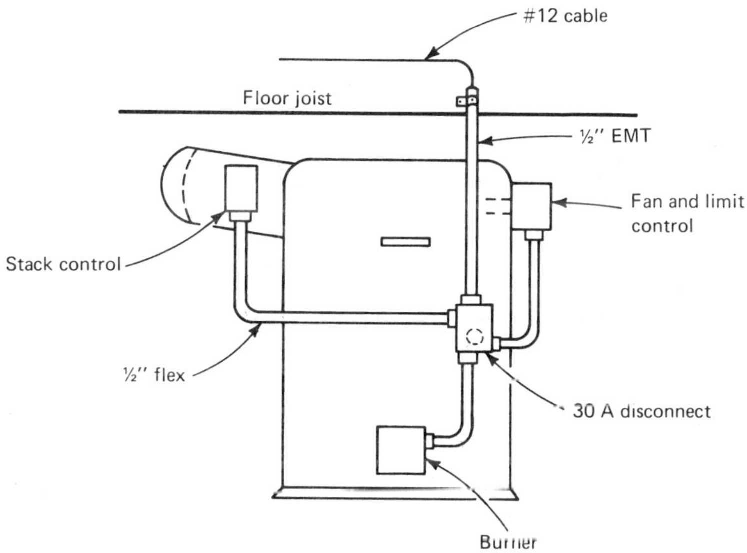

Figure 29: *Wiring for Oil-fired Warm Air Furnace*

control starts the fan circulating once the bonnet temperature rises to its setting and shuts down burner operation if a second high-temperature setting is exceeded. The warm air fan motor is internal to the furnace, and the flex to it leaves the back of the disconnect. Finally, low-voltage thermostat wiring normally extends from the stack control to the thermostat upstairs in the occupied spaces.

The components might be set in place by the heating contractor, but wiring should be performed by the electrical contractor, according to manufacturers' wiring diagrams.

Although our circuit is 115 volts, both poles of the 30-ampere safety switch may be fused down to levels required by Article 430 of the Code, which calls for motor overcurrent protection not in excess of 175% of motor full load current when time-delay fuses are used. For a ⅓-hp burner motor drawing 7.2 amperes, this would establish a maximum protection of 12.6 amperes, and for a ¼-hp fan motor drawing 5.8 amperes, this would call for a maximum of 10 amperes. The #12 wiring would terminate on the incoming terminals of the disconnect, but #14 single conductor could be extended through the flex for the rest of the wiring, which would be protected well below the level of 15 amperes. Manufacturers of fuses offer many sizes of fuses below the 15-ampere level, and we would select the closest to our maximums that did not exceed them.

A hot water system would be quite similar, except that the fan and limit control would be replaced by a combination aquastat and limit control and the fan motor by a hot water circulator motor.

Conventional gas-fired furnaces do not require a burner motor, but instead have a gas-line solenoid valve that opens and closes in response to the space thermostat, subject to being overridden by our limit section of the fan and limit control or flame detection activity of our stack control. Note that many older gas burners feature a pilot light instead of electric ignition, for burner light-off.

The majority of warm air systems initiate burner ignition upon a fall in space temperature, after which the warm air fan comes on under control of our fan switch in the combination fan and limit control as the bonnet temperature rises. When space temperatures satisfy a predetermined setting, burner operation ceases, but fan operation continues until sufficient heat dissipation has occurred and our fan switch opens to stop the fan.

Hot water systems generally ignite (1) after the water circulator has lowered jacket temperature to the point where the aquastat senses a demand for burner operation in order to restore water temperatures or (2) simultaneously with circulator startup in anticipation of falling water temperature.

Thus, in a forced hot water system, the space thermostat controls the circulator rather than the burner. Many older gravity hot water systems made use of differential water weights within the piping system. When boiler water was heated, it was displaced by heavier, colder returning water which fell by gravity through the return piping and allowed the warmer, lighter water to rise and take its place in the cast iron radiators. In such systems, the thermostat controlled the boiler burner.

HUMIDIFIERS

Exceptionally dry conditions can exist in the home in the wintertime when the heating system is hard at work keeping the household warm. As the temperature falls outside, the dew point of the air is lowered as moisture in the air is condensed out in the form of dew or frost. At these lower outside air temperatures, the capacity of the air to hold moisture is diminished. But when this cold air finds its way into our homes and is heated, its capacity to hold moisture is raised. If no additional moisture is introduced back into it, the air becomes very dry. Its relative humidity is said to be quite low. We see the effects of low relative humidity in the home in wintertime when furniture dries out and rungs of wooden chairs begin to pull out, and when the hardwood floors seem to have slightly larger cracks in them than before, and squeaks can be heard in the floor as we walk over it. Static electricity discharges become apparent when we walk across a rug and reach for a metal door handle, because the drier environment is more conducive to their occurrence than is a more usual level of relative humidity.

Our respiratory system is affected by the dryness in the air, too: when relative humidity is too low, nasal passages are robbed of their natural moistness by the arid condition of the air we breathe. Fabrics tend to rip more easily also, as their fibres tend to lose their resiliency when deprived of their normal moisture content. So maintenance of proper and comfortable relative humidity becomes both practical and advisable.

In examining warm air heating systems, we note that the environmental air within the home is drawn back to the furnace, where it is warmed by contact with a heat exchanger in which energy is consumed. At this point, and before this air is delivered back to the conditioned spaces, its capacity to hold moisture is at its peak, because of its elevated temperature. A humidifier, located so as to impart moisture to the heated air just as it comes off the heated surfaces of the combustion chamber or manifold, can effectively replenish much of the moisture that we require the air to have for our daily comfort. Hence, the regular use of furnace humidifiers. Let us now look at the wiring associated with several types of furnace humidifiers commonly found in the home today.

There are three considerations concerning electrical wiring in connection with furnace humidifiers: power, interlock, and control. Power, of course, has to do with the unit's need for electricity to operate. Interlock relates to the need to shut down the humidifying action whenever the system fan is not operating in the heating mode. And control has to do with maintenance of proper levels of relative humidity rather than permit constant operation of the humidifier regardless of the need for additional humidity. Control maintains levels of relative humidity high enough to preclude discomfort, but low enough to prevent condensation on windows in the home.

Humidifiers can perhaps be divided into two broad classes: those that accomplish their work through atomization, and those that do it through evaporation. The first kind might prepare the water for easy entrainment in the warm air stream by breaking it down into very fine droplets or mist that can be readily absorbed by the warm, dry air. The second might convert the water to a vapor through the application of heat, or merely offer a wetted medium to the passing air, allowing the hot air to draw off moisture in a repetitious process, as long as the demand for additional humidity is there. Since the evaporative type of humidifier is generally accepted as being more prevalent in residential use, we will concentrate on the wiring associated with it.

Any convenient 115-volt source can be used to power the average evaporative passive type humidifier, as its power requirements are low. The power might be derived from a nearby receptacle that could have its circuit extended to serve the humidifier. The receptacle would be connected through a relay whose contacts close only when the fan is energized in the heating mode. One way to accomplish this is to use a duct-mounted thermostat set at 90° and wired in series with the relay's contacts. Warm air would have to be present in the discharge duct and during fan operation, so that humidification could not occur during the cooling cycle, should our warm air system feature central air conditioning as well. Furthermore, a humidistat located in the return air or in the occupied space would permit humidifier operation only when space conditions warranted it.

Figure 30 shows how the humidifier would be installed in the warm air discharge duct coming off the furnace. A revolving wick or medium passes through the water reservoir in the bottom of the humidifier and is then ex-

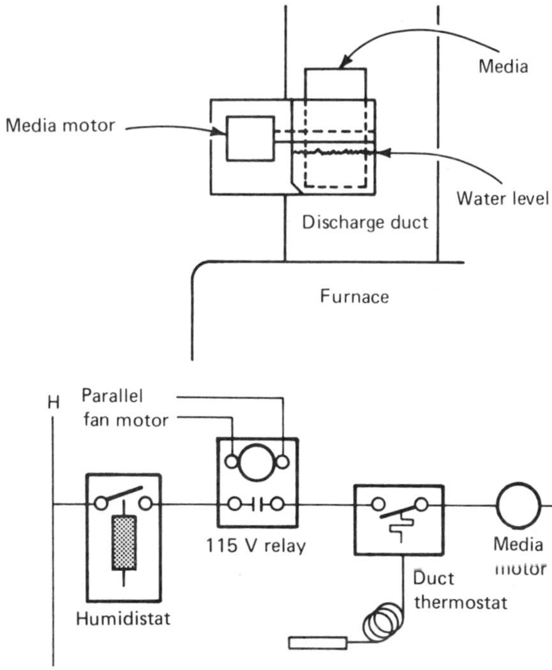

Figure 30: *Evaporative Type of Furnace Humidifier*

posed to the warm air stream. A built-in float maintains sufficient water for saturation. The medium motor is energized upon a signal for humidification whenever the fan is running and the air is warm. The wiring schematic shows the use of a separate feed, connected through the humidistat, relay, and thermostat. The relay could be omitted if the switched leg to the furnace fan were used to power the humidifier medium motor, because no hot would exist at the humidistat until the fan motor was indexed to run. Interlocking, in that case, would be through use of the fan circuit itself simultaneously with the humidifier; in effect, we would be putting the fan motor in parallel with the humidifier rather than using an interlocking relay.

Many plate-type humidifiers are in use in today's homes also. Frequently, they are nonelectric, requiring only a ¼-inch water line to be run to a self-contained metering valve built into the humidifier. The plates are made of an absorbent material and are arranged vertically in a holding rack, with their lower portions immersed in water. As the water in the reservoir is absorbed up into the medium, a level-actuated water fill valve is opened to introduce sufficient water to restore the operating level to normal. The plates are very often T-shaped, so that their upper portions overhang the water reservoir, insuring maximum exposure to the passing air stream. When the plates are mounted in the discharge plenum of the furnace, air flow is from the bottom of the humidifier toward the top. When they are mounted in a horizontal duct, air flow approaches the unit from the side and sweeps through the plates horizontally.

The plate-type humidifier is most often in operation when the furnace is heating. Humidification capacities are lower than with the revolving medium type, so interlock may not be necessary. However, plate-type humidifiers may be easily modified electrically to permit both interlock and control of their output. A solenoid control valve may be installed in the ¼-inch line to the humidifier and wired to be energized through a humidistat and warm air thermostat, using the fan circuit as the power source. For the evaporative-type humidifier of Figure 30 that utilizes a revolving medium, the medium motor would merely be replaced with the solenoid water valve and level switch. The valve would open depending upon (1) a need for an increase in relative humidity, (2) the presence of warm air coming off the furnace, and (3) operation of the furnace fan and low level in the reservoir. The voltage of the solenoid must match that of the warm air fan (115 volts in the example) as in the majority of cases. The solenoid valve would normally be closed, i.e., it would be designed to spring-return to closed to prevent addition of water when de-energized. When energized, it would permit water to flow into the humidifier if the level-actuated metering switch called for more. The absorbent plates in this type of humidifier are capable of replacement, preferably once a year, prior to the heating season.

Electrically heated humidifcation units will probably require a special-purpose circuit to power their heating elements. A 1,000-watt unit would draw almost 9 amperes on a 115-volt circuit, so a minimum 15-ampere circuit would be required. In this installation, the humidistat would energize the heating element in the unit upon demand for additional humidification through a control relay, subject to fan operation in the heating mode. A constant water level would be maintained by built-in level controls, and a safety switch would prevent operation unless the water was up to a predetermined level. The water temperature would be elevated by the heating element to force evaporation of the water in the air stream.

With all of the types of humidifiers discussed, a manual on-off switch conveniently located near the humidifier would permit manual shutdown for any desired length of time. This would interrupt all power to the unit, overriding all controls and preventing its operation completely.

NIGHT SETBACK THERMOSTAT

Some fuel savings can be effected through the use of a night setback thermostat, which has the ability to automatically index itself to a predetermined lower control point at a specified time in the evening. With fuel costs at such high levels, it makes sense to investigate the suitability of your heating system to night setback control. Cooler household temperatures during normal sleep hours can prove to be more comfortable, too. Most humidity-generating activity, e.g., cooking, bathing, and doing laundry, does not occur indoors at these later hours, so the drying effects of heating the air need to be offset to prevent arid nighttime conditions. For homes without humidifiers, lowering the temperature setting of the thermostat results in air somewhat less dry than air that would exist at higher temperatures.

The automatic night setback thermostat has been around for a long time, but renewed interest in its application has arisen due to the emphasis placed on wiser fuel management because of modern-day costs. New developments in thermostat design have contributed to wider utilization, carrying over into homes where simple two-wire control exists and making unnecessary the use of a thermostat that requires both sides of the control circuit to be located at the thermostat for operation of the clock motor. It is thus appropriate to examine some of the alternatives available to today's home owner who wants to apply reduced temperature control to his home at night for system economy and greater comfort.

Historically, fossil-fuel-fired heating systems have been particularly suitable for night setback, and because such applications are quite abundant today, we will look into typical system control configurations that demonstrate how the setback is accomplished in such systems.

An oil-fired warm air system generally has the burner cycled by the thermostat, so that falling space temperatures can initiate a call for heat, warm the furnace plenum, and cause the combination warm air control and limit to effect fan start-up. An anticipator located within the thermostat applies false heat to warp the bimetal prematurely, causing burner shutdown just before space temperatures coincide with the setting of the thermostat. In many applications like this, three control wires from the furnace stack switch control are extended to the location of the thermostat. Control circuit voltage is usually 24 volts and 3-wire low-voltage cable connects the thermostat to the stack switch control at the furnace. Figure 31 shows the

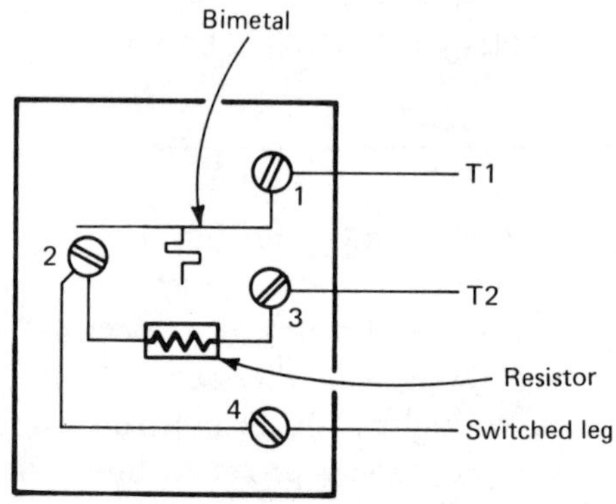

Figure 31: *Two-position Heating Thermostat with Anticipator*

thermostat with its three wires and the built-in anticipator: T1 of the control transformer, T2, and the switched leg. A fall in temperature causes the circuit to be completed from 1 to 2 within the thermostat, placing voltage across the resistor, which reacts by becoming warm. At the same time, voltage flows back to the furnace controls on the switched leg. A small clock motor could be connected across terminals 1 and 3 of the thermostat, because they are always "hot." Then the clock could operate cams or trippers to swap control from a "day" bimetal to a "night" bimetal at preset times each evening, and return control to the "day" bimetal at a preset time each morning. Technique may vary from one manufacturer to the other, but from a performance standpoint, this is essentially what happens in the night setback thermostat. Home occupants determine the times at which evening and morning changeover occur by adjusting the night setback or clock thermostat according to their desires. Changes are easily made, so complete flexibility can be exercised in personal choice of times at which temperature change is to occur.

Figure 32 illustrates the internal wiring of a basic night setback thermostat. Note the two bimetals. The top one is the day bimetal, and the lower one is the night bimetal. The day bimetal will make contact on a fall in temperature before the night one. There may be five degrees or more (adjustable) in the different control points of each bimetal. We take advantage of this differential by setting a clock-operated contact between the load-side terminals of the bimetals. The clock closes the contact during the day and opens the contact during the night. Therefore, the day bimetal can start and stop the burner during daytime hours, but the night bimetal must close before heating equipment can be energized during the sleeping hours. We still see the resistor present for heat anticipation during the day or night to prevent system temperature overshoot.

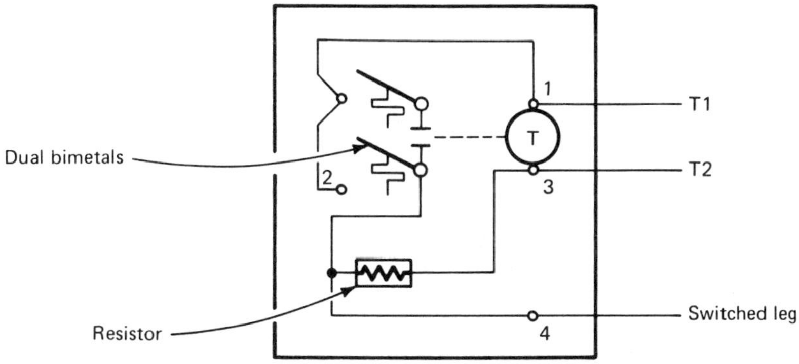

Figure 32: *Two-position Heating Thermostat with Anticipator and Night Setback*

Invariably, a 12-hour clock is part of the thermostat for easy readout of the time of day, giving us the convenience of an electric clock as an integral part of the thermostat to facilitate temperature setback decisions. Our view of the night setback thermostat is limited to the electric side of the thermostat's internals, and the mechanics of temperature control selection and actual mechanical setback orientation are not shown.

Two-wire-control heating systems have historically been denied the benefits of reduced nighttime temperature control, but recent developments have introduced night setback thermostats that can utilize the switched leg back to the furnace for restoration of an internal power source for powering the time clock motor. Such restoration occurs during the shutdown period of the heating plant, when the switched leg is utilized for trickle current flow to revitalize the thermostat's self-contained internal power supply. Normal furnace-control components do not react to the tiny, limited current passing through them during the recharging period, as the controls themselves become mere conductors for passage of the trickle charging current back to T2 of the control circuit transformer. During a call for heating, no voltage drop occurs over the switched leg, because contacts close, impressing control circuit voltage directly onto furnace controls that energize burner control relays. But in the absence of a call for heating, when the thermostat's internal charging components are placed in series with furnace components that require full voltage to react, the furnace control relays drop out, and their holding coils become low-current conductors for the recharging of the thermostat's self-contained power supply.

Figure 33 shows how a two-wire setback thermostat might work. Our same two bimetals are present, but their T-1 connection parallels an AC/DC converter and charging module. Whenever the heating contacts are open, current flows through the charging module, using the switched leg as the path

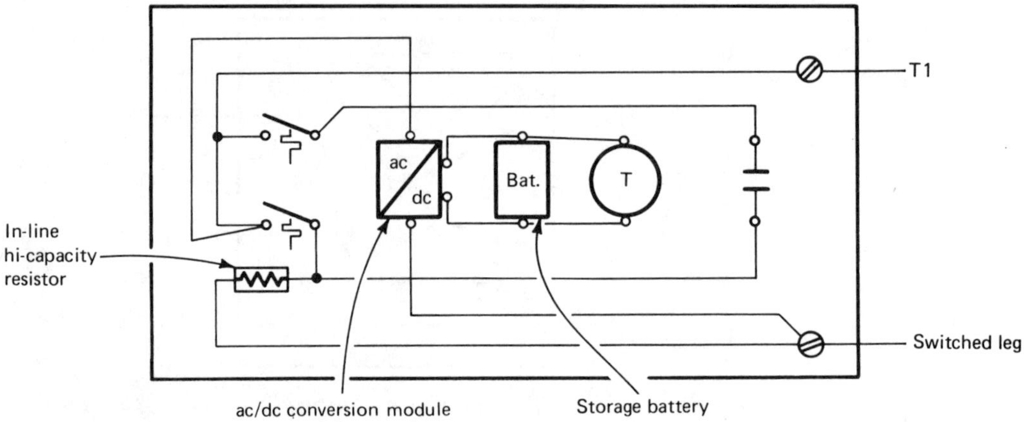

Figure 33: *Two-wire, Two-position Heating Thermostat with Anticipator and Night Setback*

back to transformer T2. When either of the heating bimetals closes its contacts, current follows the path of least resistance and bypasses the charging unit. Furnace heating relays pull in, and our system-sized in-line resistor, selected to handle full current flow to the control relays, continues to serve its role of biasing thermal elements slightly to effect even temperatures in the spaces being served.

As a homeowner, if you decide to purchase one of the several different types of night setback thermostats for use in your home, be sure you match the voltage characteristics of the one you intend to purchase with your system's control voltage. This is particularly important for operation of the clock. Undoubtedly, the control circuit voltage will be 24 volts, as this is pretty standard. Follow manufacturer's directions and wiring diagrams, and if you have any doubts, be sure to consult a qualified serviceman or electrician. A sufficient number of different models of thermostats should be available in the marketplace to prevent the necessity of pulling new wiring from the furnace to the thermostat. All that should be necessary is removal of the old thermostat and installation of the new, if proper care is exercised in selection of the new night-setback thermostat.

Lower temperatures in the home at night should lessen the problem of "sweaty" windows in cold weather, because the capacity of the cooler night air to hold moisture is reduced. Higher relative humidity results, with more comfortable sleeping.

If you are experiencing short-cycling of the burner on your warm air system, examine the thermostat for the presence of an adjustment lever marked "shorter-longer." This refers to burner cycle and allows your thermostat to be adjusted to match current draw of the coil in the burner control relay. The numbers refer to the amperage rating of the coil. Realizing that most homeowners do not get involved with the technicalities of the heating system, thermostat manufacturers have marked the adjustment lever "shorter-longer" to facilitate adjustment without having to have knowledge of coil rating. Moving the level to "longer" compensates for burner short-cycling by permitting the greater current draw to generate less false heat within the anticipator. This should lengthen burner cycles, as the thermostat is slower to react to premature shutdown. If, even with the adjustment at the maximum "longer" setting, short cycling still occurs, it might be that recent furnace repair work saw the installation of a primary control whose coil current draw is greater than the old one, exceeding thermostat adjustment compensation capabilities. If stack switch primary controls are no longer available with coil capacities that match your thermostat, the thermostat could be replaced or a smaller relay that is designed for lighter duty with a lighter coil could be used to pilot the burner relay. Such pilot relay selection should incorporate an attempt to match your thermostat's anticipator adjustment range scale. With a successful match, you can obtain proper burner cycle lengths, since the smaller relay's coil is what the thermostat energizes when it calls for heat. In such an arrangement, the wiring that would otherwise have gone to the thermostat's heating contact would go to the contact of the pilot relay instead. 24 volts would have to be picked up to serve the smaller relay, and this would probably require a separate 115- to 24-volt transformer, as the internal transformer within the stack switch is usually sized for internal loading only or to handle only the relay coils within the stack switch itself. Figure 34 illustrates the situation schematically. Note how the low voltage circuit out of the stack switch that originally was routed to the thermostat is

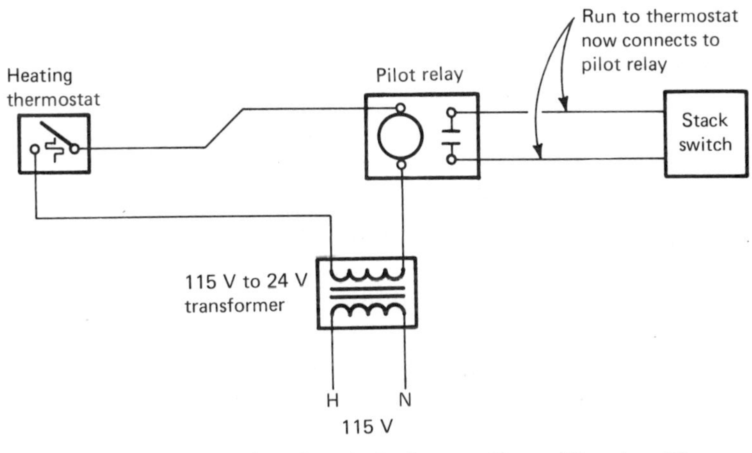

Figure 34: *Pilot Relay for Anticipator-Type Heating Thermostat*

now connected into the contact of the pilot relay. This lighter-duty relay is selected to compliment the current rating adjustment of the thermostat's anticipator.

On the other side of the coin, burner cycling may be too long. Although this is not encountered too often, it could arise in the event the burner control relay in the stack switch did not load the anticipator circuit enough to effect resistor warm-up for proper anticipation. If, with the lever adjustment moved to the extreme position in the "shorter" direction, burner cycling does not decrease, dummy loading of the control circuit could be employed to bring the current draw within the range of the thermostat's adjustment. This would consist of connecting a light-duty relay in parallel with the burner control relay in the stack switch to increase the amount of current in the thermostat's control circuit when it calls for heat. Here again, however, an external 115-to-24-volt transformer might have to be used for the additional power requirements necessitated by the second relay coil. This condition is not as prevalent as burner short cycling, since thermostat anticipator adjustments usually are designed to embrace the range of currents found in the lighter type of relays anyway, e.g., the relay used as an integral part of the stack switch control unit.

Thermostat anticipator adjustments should be made in small increments and the burner allowed to operate for a day or so at the new setting until the operation can be observed and adequately appraised. Then, the additional amount of correction in adjustment setting can be gauged more closely, and the second lever adjustment will fine-tune the setpoint to bring it more closely in line with the exact setting required.

CENTRAL AIR

Many of the warm air heating systems in existence today are candidates for modification to central air conditioning systems. The presence of the furnace fan and associated ductwork constitutes a major ingredient in the recipe for cooling from a central fan source. Most often, the addition of a cooling coil, or evaporator, in the discharge of the furnace and installation of a remote condensing unit with associated wiring and controls is all that is necessary to expand the services of the central plant to provide cooling as well as heating. Insulation might have to be applied too, to metal ductwork where sweating could occur, in order to prevent moisture problems attributable to condensation of humid air on the colder duct surfaces.

On a call for cooling from the heating-cooling thermostat in the occupied spaces, operation of the condensing unit begins simultaneously with start-up of the indoor circulation fan or furnace fan. As refrigerant is released through an expansion or metering valve into the lower-pressure environment of the evaporator or cooling coil, it absorbs a considerable amount of heat as it flashes into a gaseous state. Surfaces of the finned cooling coil come in contact with the passing conditioned air stream and extract significant amounts of heat and chill the air. As the air reaches its dew point, condensation forms on the coil and is drained off. The refrigerant gas is then drawn back to the outdoor condensing unit, where it is pressurized by the compressor before being circuited through a condensing coil under high pressure. As the pressure and cooling effects of the condensing coil combine to change the state of the refrigerant gas, it condenses back to a refrigerant liquid, where it is maintained under pressure in this state until it reaches the point in the refrigerant circuit where it is again released into the evaporator. Thus, the unique properties of modern-day refrigerants are utilized in this compression-expansion process to remove heat from indoor spaces and eject it to the outdoors in an air-to-air heat exchange process, with the refrigerant being the intermediary in the exchange.

We now take a look at the power and control wiring typically associated with installation of central air conditioning into existing warm air fan systems. The furnace circuit continues to be used for powering the indoor blower during cooling, but provision must be made for initiating fan operation by paralleling the contacts of our fan switch in the combination fan-and-limit control.

In Figure 35, a heating-cooling room thermostat is shown with its internal selector switches and thermally actuated, single-pole double-throw contact. Note the anticipator in the heating circuit. This particular type of thermostat requires four low-voltage wires, so existing 2-wire or 3-wire cable to a heating-only thermostat will have to be replaced. Terminals on the thermostat are labeled "R," "Y," "W," and "G." Our 24-volt control voltage, derived from the condensing unit control circuit transformer, is shown originating from low-voltage terminal points CT1 and CT2. One side of the control circuit is connected directly to R on our thermostat. Y connects to cooling relay CR in the condensing unit outside; W connects to heating relay HR located inside near the furnace; and G is connected to fan relay FR. Thermostat selector switches S-1 and S-2 allow mode indexing to dictate central system performance. System switch S-1 offers three positions; *cool*, *off*, and *heat*. In *cool*, heating is locked out, and a rise in temperature above the thermostat's setpoint brings on the furnace circulating fan and the outside

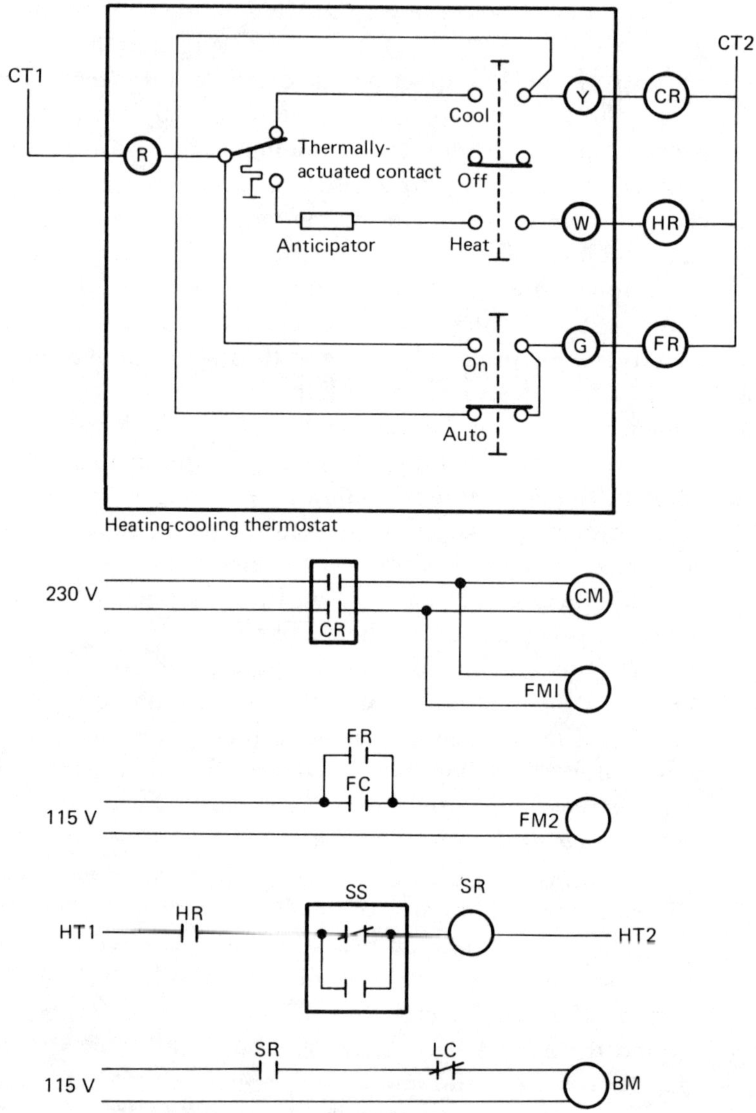

Figure 35: *Thermostat with Control Wiring Oil-fired Warm Air Heating With Added Cooling*

condensing unit. In *off*, the system is completely shut down. In *heat*, cooling is locked out, and a fall in temperature below the thermostat's setpoint brings on the burner, followed by fan operation upon a rise in bonnet temperature at the furnace. Fan switch S-2 offers two positions: *on* and *auto*. In *on*, the furnace circulator fan runs continuously for constant movement of air throughout the house. In *auto*, the furnace fan comes on simultaneously with start-up of cooling, or after plenum warm up in heating.

Compressor motor CM and condenser fan motor FM_1 are started and stopped through the double-pole, normally open contacts of cooling relay CR. These contacts are load-rated, and when closed, impress 230 volts upon both motors in response to a demand for cooling.

Furnace fan motor FM_2 is controlled during the heating mode by fan control FC, which closes its contact upon a rise in bonnet temperature after the operation of the burner has warmed the heating chamber. However, the contact of fan relay FR is wired in parallel so that indexing of switches for continuous operation or a demand for cooling can initiate fan operation as well.

Stack relay SR, an integral part of the stack control that is positioned to sense flue temperature for verifying burner operation, is energized whenever heating relay HR closes. The normally closed timed contact in stack switch SS opens after a short duration, but not before the normally open, thermally actuated contact closes to maintain the circuit. In this manner, the presence of flame is confirmed, and relay SR is permitted to stay energized. Should flame not be present, stack relay SR will drop out, because closure of the normally open thermally actuated contact will not occur due to the absence of a rise in flue temperature.

Burner motor BM begins to operate through the stack relay contact SR and continues as long as the contact is closed, subject to shutdown by the limit control's normally closed contact LC, which will open if bonnet temperature rises to too high a level. Actually, the limit control LC and fan control FC are combined in the same unit, which is known as a combination fan and limit control, and is installed in the discharge plenum or bonnet of the furnace to sense air temperatures coming off the furnace.

Furnace fan motor FM_2 and burner motor BM are fed by the same 115-volt furnace circuit but will likely be fused separately in a 2-pole fused safety switch mounted at the furnace.

Control circuit voltage for the stack relay SR is 24 volts and is derived from the terminals labeled HT1 and HT2 of a transformer built into the stack switch. In the original heating-only control circuit for our furnace, SR would have been controlled directly from the thermostat. But in the heating-cooling control circuit, heating relay HR has its contact wired into the low-voltage furnace burner stack relay control circuit in place of direct wiring to the thermostat. This is because we have extended just one low-voltage source to the thermostat and it originated at the condensing unit. The thermostat we selected makes provision for only one control circuit being introduced into it, so isolation of the furnace's burner control circuit becomes necessary. This is accomplished with the 24-volt heating relay HR.

The heating and fan relay could be located at the furnace in a suitable enclosure. Then the fan motor circuit that powers FM_2 could be extended to our relay cabinet or enclosure, perhaps through flexible metal conduit if the run is less than 6 feet. And the low voltage lead from our stack switch that used to go upstairs to the old heating-only thermostat could also be routed into the relay cabinet, for connection to HR's contact. Of course, a new power circuit of 230 volts at perhaps 30- to 40-ampere capacity would have to be run to the outside condensing unit through a weatherproof serviceman's disconnect. Then, weatherproof flexible conduit would bridge over to the condensing unit from the disconnect. If the flexible conduit is sized large enough, the low-voltage circuit for the cooling relay CR that is inside the condensing unit could be run into the unit together with our power conductors.

Actually, the wiring scheme in Figure 35 is relatively simple and fairly basic to all combination systems in which warm air furnace heating systems have been modified through the installation of central air conditioning.

YEAR-ROUND CONTINUOUS SPACE CONDITIONING

In some of the larger homes being constructed today, the trend is toward stationary sashes, where many of the windows in the home do not open. This allows a permanent seal around the window against outdoor air temperatures, and it may also save first-cost dollars over sashes of the conventional type, which are movable and must be weather-stripped. Infiltration of colder air temperatures in the wintertime is reduced by the permanent barrier afforded by these fixed windows, and better energy management may be obtained through reduced fuel or power consumption, attributable to the more effective ban against air leakage around windows. There is greater security against intruders also, in windows which cannot be forced open. And where such windows are installed, special considerations are frequently given to space conditioning inside the home, to insure constant temperatures that maintain comfort standards without the necessity of opening windows. Providing sufficient air movement within the home at the correct temperature becomes the job of the heating and air conditioning system, just the same as in conventional homes except for one difference: economy needs to be introduced into the system's operation to forestall unnecessary energizing of mechanical heating or cooling. This economy can be achieved through the utilization of energy-saving thermostats and the introduction of fresh air through the space conditioning system if the fresh air temperature is suitable for cooling interior spaces. We need to turn our attention, then, to these air-handling units that might be applied in such a situation and consider the electrical controls pertinent to creating the indoor environment desired at the lowest possible cost.

An air-handling unit is any fan system that treats the air that flows through it in such manner as to make it warmer or cooler and/or cleans it through filtering, ultimately delivering it back to the interior spaces to effect temperature control. Long prevalent in the world of commercial temperature control, air-handling units, in smaller sizes, are finding occasional but growing application in residences of some prestige, where the first-cost of installation of the additional control equipment is not very significant when compared to the overall project cost. Perhaps the most noteworthy feature of the kind of air handler under consideration here is its ability to make use of outside air as a cooling medium whenever (1) cooling of the interior spaces is required, and (2) the temperature of the outdoor air renders it suitable for cooling. To get at the outside air, a duct must be extended from an outside air intake point to the suction-side plenum of the air handler and properly installed with dampers to permit selection and control. Additionally, a lateral needs to be extended from the return air duct to the outside or into a ventilated attic, so that returning air can be discarded when outside fresh air is being admitted.

To accomplish all this, automatic control dampers will be installed in the ductwork at the unit. These dampers are shutters or louvers that can be opened or closed to control air flow into the unit. They are installed on the suction side of the air handler and controlled by actuators linked to drive shafts. These electric actuators are operated by thermostats in the mixing plenum, which either regulate the dampers in a two-position mode or modulate them for more gradual action. We will examine some typical control schemes that may be applied to air-handling unit intake dampers. First, let us look at the unit.

In Figure 36, we visualize the unit hung from the rafters in the attic space. The fresh air inlet is through the wall at the gable end, and the fresh air damper can be seen linked to the same actuator, M1, that drives the return air and relief damper through linkage. Whenever the space temperatures in the occupied areas below would demand cooling, a rise in temperature at the room thermostat would first de-energize all heating. Since fan operation in the air handler is continuous, on a continued rise in room temperature the fresh air damper would be opened fully while the return air damper would

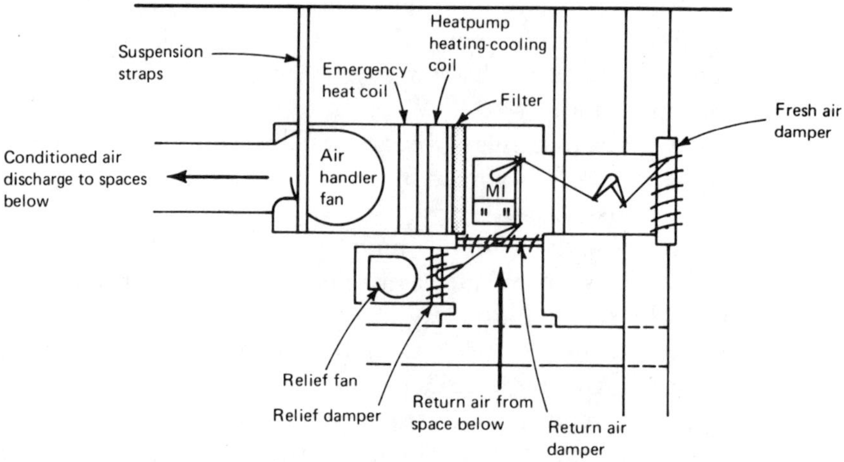

Figure 36: *Attic-mounted Residential All-electric Air Handling Unit*

be closed and the relief damper opened. At the same time, our auxiliary relief fan would start and exhaust returning air from the spaces below into the unoccupied attic area, whereupon the discarded air would escape out through stationary louver vents high in the gable ends. Starting of the auxiliary relief fan is necessary, because during this ventilation period the main air handling unit fan is isolated from the ceiling return. In the event outside air temperatures were too warm for use in the free cooling, ventilation cycle, the dampers would remain closed to outside air, and space temperatures would be allowed to rise slightly during a deadband span in the room thermostat for energy conservation, until, at probably 78° space temperature, mechanical cooling would be energized. If the introduction of outside air during the ventilation period did not satisfy the room thermostat, and space temperatures continued to rise, mechanical cooling would be energized anyway, as the next step in the cooling sequence. Fresh air would continue to be introduced as long as its temperature was below a predetermined cutoff setting. Should outside air temperatures rise above the cutoff point, the dampers would revert back to their original status, closed to outside air and relief, and opened for 100% returning air. The auxiliary relief fan would then shut down, also. Thus, this ventilation cycle is utilized in an energy-efficient mode to introduce free cooling as the first stage of cooling when outside air temperatures are effective. New energy-saving, no-activity deadband thermostats would call upon the air handler's ventilation cycle for this free cooling between the points where heating and cooling are energized, in order to forestall start-up of mechanical cooling for as long as possible, so long as cooling was called for and outside air temperatures rendered it suitable.

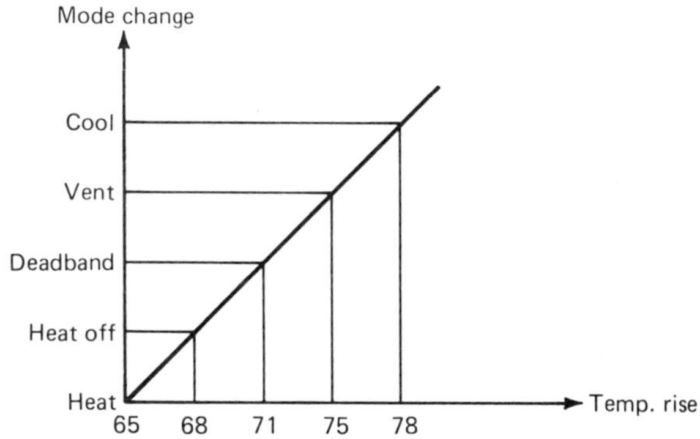

Figure 37: *Graph of Space Temperature vs. Unit Mode*

Figure 37 shows the reaction of the air-handling unit to a rise in space temperature. In accordance with suggested settings for reduced energy usage, our heat cuts off on a rise to 68°. No further energy is expended until 78°, when mechanical cooling is utilized. If the outside air temperature is 50°, we might very well satisfy space temperature demands for cooling through the exclusive use of outside air without bringing on mechanical cooling. It is not unusual, even in cool weather, for a need for cooling to exist where many people are assembled together, as in the case of a party. In such a situation, the introduction of outside air would handle the cooling until the ventilation cycle is shut down upon a fall in temperature.

As a further step toward energy conservation, space temperatures could be setback 5° at night and whenever the house was empty during the winter to a 60° setpoint, by stopping the air-handling unit fan with a time clock. Then, when the space temperature fell to 60°, a separate night thermostat might restart the unit in the heating mode to prevent temperatures from falling below 60°. The unit would be cycled on and off as required during this setback period, and then restarted for continuous operation during the "day" or "occupied" mode for restoration of normal temperature levels.

In Figure 38, illustrating the fan and ventilation controls only, we note that the 115-volt power circuit to the air-handling unit fan motor is broken through a time clock contact for starting and stopping of the fan according to a predetermined schedule. Both an override switch and night thermostat parallel this clock contact so that the unit fan can be started for uninterrupted operation with the override and cycled on when space temperature falls to 60° by the night thermostat. Wired parallel to the main fan motor is the feed for our damper actuator, so that it may be activated by our low-voltage control circuit, provided the time clock contact is closed to put power on the actuator. When the unit fan is running, the vent relay may be energized from the room thermostat, and if the outside air temperature is below 60°, the outside air thermostat closes its contact to complete the control circuit to the damper actuator, which will then open the fresh air and relief dampers and close the return.

The night thermostat would be located on the same wall and adjacent to the day thermostat, but would be set 5° lower. Then, when the night thermostat was showing a cold enough temperature to start the fan for heating in the winter, the day thermostat would be showing an even colder temperature, with respect to its higher setting, and the unit would start on full heat.

Whenever power is cut off to stop the unit fan, the damper actuator will close off to outside air because of its "spring return" construction. In this connection, the fresh air damper is said to be normally closed, as is the relief damper, while the return air damper is said to be normally open. These terms refer to the state of the dampers in their normal or de-energized position. Assuring cutoff of outside air whenever the fan stops prevents the migration of cold outside air into the unit, and possibly into the house, in the winter. The vent fan is shut down simultaneously with relief damper closing and started simultaneously with relief damper opening through the contact on the relief fan relay. Therefore, the outside air damper and the relief damper and relief fan are all activated simultaneously, for an interlocked mode of operation. Additionally, stopping the main unit fan removes power through the relief fan relay contact, assuring shutdown of the relief fan anytime the main air-handling unit fan is stopped. Thus, the relief fan is subject to a call for operation only when outside temperatures fall below 60° and the main fan is in operation.

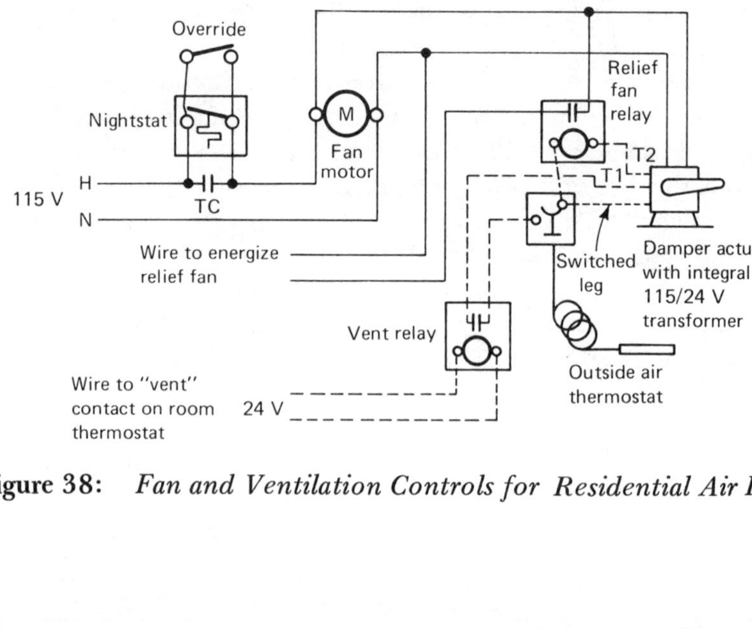

Figure 38: *Fan and Ventilation Controls for Residential Air Handler*

Were we to have selected modulating controls to serve our ventilation cycle, our damper actuator's response would be gradual and proportional to the deviation from setpoint at the room thermostat. Instead of relays, end switches, mounted on the actuator and operated by the same drive shaft that positions damper crankarms, would close cam-driven contacts to complete electrical circuits to energize the relief fan. Or we might opt for a mixed air controller to inaugurate the ventilation cycle—one that could modulate the intake dampers based on mixed air temperature in order to maintain its set-point of, say, 55° or 60° whenever the vent contact of a conventional room thermostat was closed. The outside air thermostat would remain as a limit control to bias the proportional control circuit for the purpose of returning dampers to their normal state whenever outside air temperatures were unsuitable for cooling. Proportional controls offer the advantages of gradual reaction to changing temperatures and the ability to create more stable levels of temperature control, and these can be quite important with respect to dampers.

A proportional controller, in many cases, reacts to temperature change by unbalancing a 3-wire circuit to energize a balancing relay within the actuator, which in turn energizes a reversible motor to drive the shaft so as to move damper crankarms. At the same time, motor motion repositions resistances within the actuator on the other end of this 3-wire circuit, to restore circuit balance and return the balancing relay to its null position, stopping the actuator motor. In this manner, actuator movement is made proportional to the amount of deviation at the controller in either of the two directions necessary.

ELECTRIC BASEBOARD HEAT

One of the most versatile forms of heat being used in homes today is electric baseboard heat, because of its unique flexibility pertaining to where and how it can be located. Existing rooms that require supplemental heat or new additions are both excellent candidates for electric baseboard radiation, which can be controlled with wall-mounted or unit-mounted thermostats.

In determining how much heat is required in a home, normal allowances of 8 watts per square foot of finished living space are usually ample. The most popular units are of the medium density class, offering 250 watts per foot of active element length and operating at 230 volts. Given a room size of 18 feet x 22 feet, we find there are 396 square feet of floor space. At

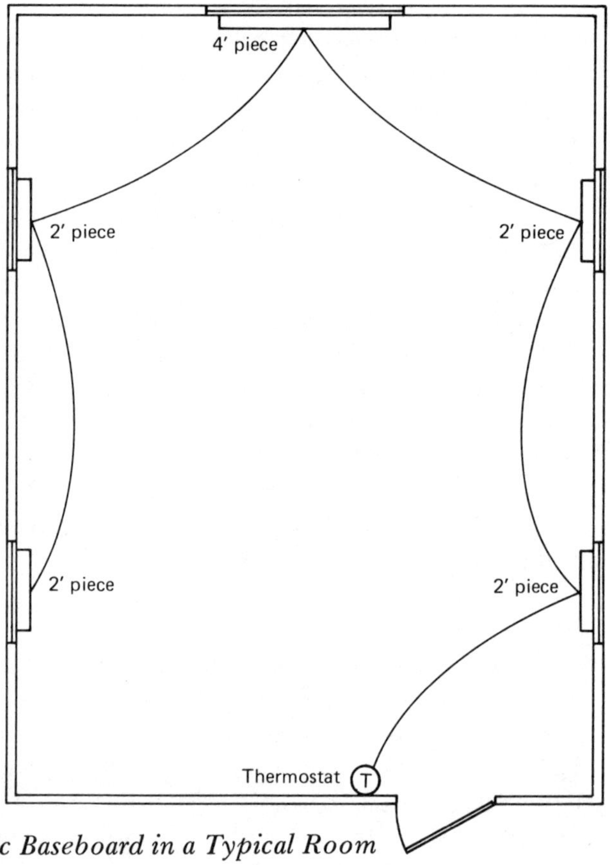

Figure 39: *Electric Baseboard in a Typical Room*

8 watts per square foot of space, 3,168 total watts of electric heat are required to insure adequate comfort under extreme outside conditions. Dividing this figure by 250 watts per foot of active element, we see that 12.6 feet of baseboard radiation is required. Normal available lengths offered by manufacturers are 2, 3, 4, 6, 8, and 10 feet. This may vary slightly from one manufacturer to another, but basically these are the sizes most often listed. Our decision as to what lengths to use must be influenced by the layout of the room and the positioning of doors and windows. Baseboard can be uniquely effective when placed under windows, as it serves the dual purpose of heating the room and offsetting the reverse air flow that occurs at windows, i.e. the downward draft immediately below the window that is due to the cooling effects of glass interacting with the room air that comes in contact with it to make the air heavier and cause it to migrate to the floor level.

Figure 39 shows a room with 5 windows, 4 of the single- and 1 of the double-sash variety. Our selection is for a total of 12 feet of baseboard, in

5 pieces to permit placement under windows. If this were an already existing room, our #12 cable could perhaps be routed through floor joists in the crawl space and the wiring to the thermostat fished into the partition void space. Care must be taken to locate the wall-mounted thermostat on an inside wall. Outside walls are somewhat cooler than the conditioned space, owing to the fact that some conduction occurs through the wall material. Thus, thermostats mounted on them may sense wall temperature, which might be falsely influenced by the outside air, instead of room temperature. Most thermostats are located on the strike side of doors to the room, above light switches.

Once the cable is fished in to the thermostat location, the hole around its entry point should be sealed to prevent stack effect from lifting cooler air from the crawl space below into the thermostat. This would cool the thermostat and cause it to maintain control at a somewhat higher point than it was set for.

The wooden baseboard trim must be cut in existing rooms to permit the electric baseboard heating element to be mounted against the wallboard or plaster wall. Measure the length of the unit's outer housing, and mark this measurement before cutting the section of shoe moulding and wooden baseboard trim to be removed. In new work in additions, elements are set after wall material is up, but *before* trim work is begun.

In new work, cable entry into the heating unit can be from the back if cables are turned out at the proper location before walls go up. In old work, rise from the crawl space below, and come in through the bottom knockout

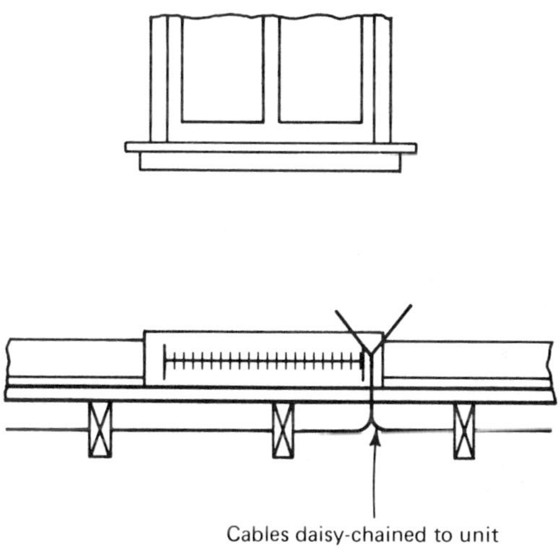

Figure 40: *Electric Baseboard Heater Rough-in*

in the baseboard heater. Use approved cable connectors, and secure the heaters with pan-head self-tapping screws into wooden studding or furring.

Figure 40 shows the open heater, with cables coming into the wiring compartment of the baseboard heater from below. Crawl-space junction boxes need not be set in this kind of arrangement, since joints are made in the wiring compartment of the unit itself.

Our power circuit for the electric heating in this room is determined by dividing total wattage of the units by the operating voltage of 230 to find amperes. Twelve feet of active element at 250 watts per foot yields 3,000 watts, and this divided by 230 volts produces 13 amperes. Electric heat is classified as a continuous load, so 125% is added, for a calculated product of 16.25 amperes. A 20-ampere #12 copper wire non-metallic sheathed cable is selected. In an existing home, 20-ampere overcurrent protection in the form of either a circuit breaker or a 20-ampere fuse, needs to be added to the panelboard for this new circuit. Hopefully, there will be sufficient space in the panelboard, and the existing service equipment will be large enough, to permit this protection.

Article 220 of the Code provides a method for making a determination whether the existing service in a home is of sufficient size to permit the installation of electric heat without the necessity of a change in service. The article offers an optional calculation for existing homes with 60-ampere service—a figure not uncommon in very old homes—to deal with considerations related to adding electric baseboard heat, as follows. The larger of air conditioning *or* the baseboard electric heat load is listed at 100%. Then, the first 8,000 watts of all other load is entered at 100%, and the balance is entered at 40%. All other load denotes 1,500 watts for each 20-ampere appliance circuit; 3 watts per square foot for lighting and portable appliances; and nameplate ratings for ranges, ovens, counter-mounted cooking units, and all other appliances (including water heater).

It is significant to note that had we selected individual wall thermostats or those mounted in the independent baseboard heater sections, and had the total number of individual sections exceeded four, our electric heat load could have been totaled in with other loads and been subject to the 100/40% consideration. This recognizes the fact that individual heat units can cycle on and off at different times under individual thermostatic control, whereas our heaters, falling under control of a single thermostat, all come on at once.

For an existing service of 100 amperes or more, a slightly different approach is tendered by the Code in Article 220. Air-conditioning load and total electric heat load are itemized at 100% and 65% respectively, and the smaller of the two noncoincidental loads is dismissed. Then, the first 10,000 watts of remaining load is entered, with the balance of the remaining load entered at 40%. Remaining load in this example is the same as "other load" in the previous one.

These examples, given in the Code for National Reference Use, might be subject to local modification, so a check with local enforcement authorities is always in order.

Let us consider thermostatic control of this electric heat. As mentioned, our choice has been for one wall-mounted thermostat to energize all elements. Electric baseboard heat manufacturers offer these wall-mounted thermostats as well as the unit-mounted variety. Their thermostats most often are of the line-voltage type, breaking the entire load when their contacts open. The mass of the sensing mechanism, being somewhat larger than that of the sensing mechanism in a low-voltage thermostat, may permit more differential to occur between "on" and "off" modes than the mass in the low-voltage one. For this reason, some of the line-voltage wall-mounted thermostats for electric baseboard heater control are marked "Cooler-Normal-Warmer" rather than with degrees, and room thermometers are not incorporated into the cover. An alternative would be to utilize low-voltage control with a combination transformer-relay unit to effect contact opening and closing. Figure 41 shows such a control unit, which could be mounted in the crawl space.

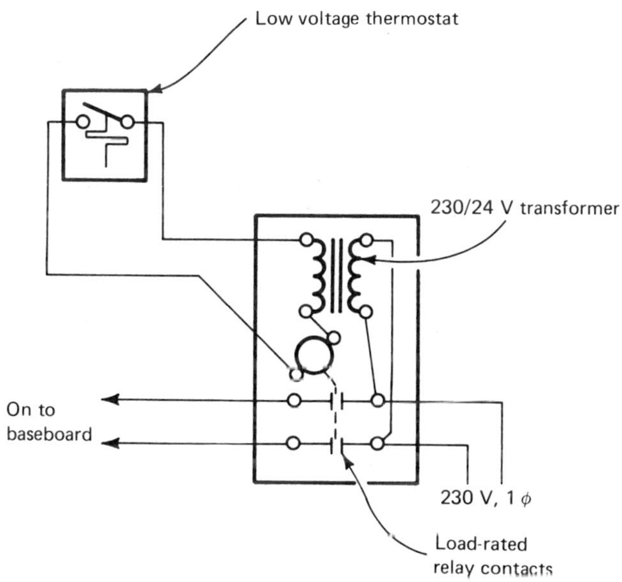

Figure 41: *Low-voltage Control of Electric Baseboard Heaters*

In the figure, we make use of a smaller, low-voltage thermostat, which offers greater sensitivity, together with an integral cover-mounted thermometer. Low voltage is derived from a transformer in the transformer-relay unit, and, on a fall in room temperature, the thermostat merely energizes the holding coil of the electric heat relay, thereby closing its contacts to permit operation of the baseboard heaters. Contacts and thermal element size of the thermostat can be small, because the only load it has to switch is that of the holding coil of the relay itself—and that at only 24 volts. Normally, a click can be heard when the relay "pulls in," and if this were objectionable, the unit could be removed to another part of the house where the occasional noise related to closing of the contacts would not matter.

Unit-mounted thermostats offer the advantage of not having to locate the thermostat on a wall, with its associated drilling and fishing in old work; and, of course, the diversity consideration for service load can be an advantage, allowable for use in our calculations when the total number of units is 4 or more. But a greater accuracy, in terms of setpoint and actual control point, can probably be attained when the thermostat is located at the "living level," where its room height is such that it can sense temperature felt by the occupants of the room, rather than at the floor level. However, many locations for electric baseboard heaters, such as entryways, halls, and foyers, might not demand the closer settings allowed by wall-mounted thermostats, and the swap-off for not having to fish and penetrate finished walls may represent the better approach.

SOLAR ENERGY IN THE HOME

Of increasing importance in considerations related to operational costs in new homes today is the harnessing of solar energy to perform some of the work historically done by conventional means. Some significantly large electrical loads can be partially replaced today through the installation of solar collectors and associated controls. Consider domestic hot water in the home of a family of five people. With the bathing, dish washing, and clothes washing that must be done, usage of hot water becomes routine, and heating water to keep up with the needs contributes sizeably to the electric bill each month. Domestic hot water is one area in which solar energy can be used to lighten the electrical load.

The conventional electric hot water heater operates on a demand basis by responding to falling tank temperatures through energizing possibly a

4,500-watt heating element. If it operates 10 hours out of the day, 45 kilowatt hours of electrical usage must be paid for, possibly at the rate of 6¢ per kilowatt hour. This adds up to $2.70 per day for hot water, or $81.00 per month. If utilization of solar energy to heat hot water reduced operation of the electric hot water heater by one half, a savings of $40.50 per month could be realized. If the 10-hour per day rate of hot water usage occurred for nine months out of the year, we could realize an annual savings of $364.50. This figure divided into the total cost of a solar energy project for heating domestic hot water would give us the pay-back period.

Sunny days are a prerequisite for successful operation of the solar energy station. So controls must be employed that can detect collector activity and effect collector shutdown on cloudy days and at night, when solar heat generation does not occur. In this manner, reverse activity is guarded against, and loss of temperature through the collector is prevented.

The flat plate type of collector is perhaps the most prevalent at this time. It can be seen in widespread use angled to face toward the sun's predominant position. A water-glycol mixture might be piped through the circuiting in the collector plates to absorb the sun's heat and conduct it back

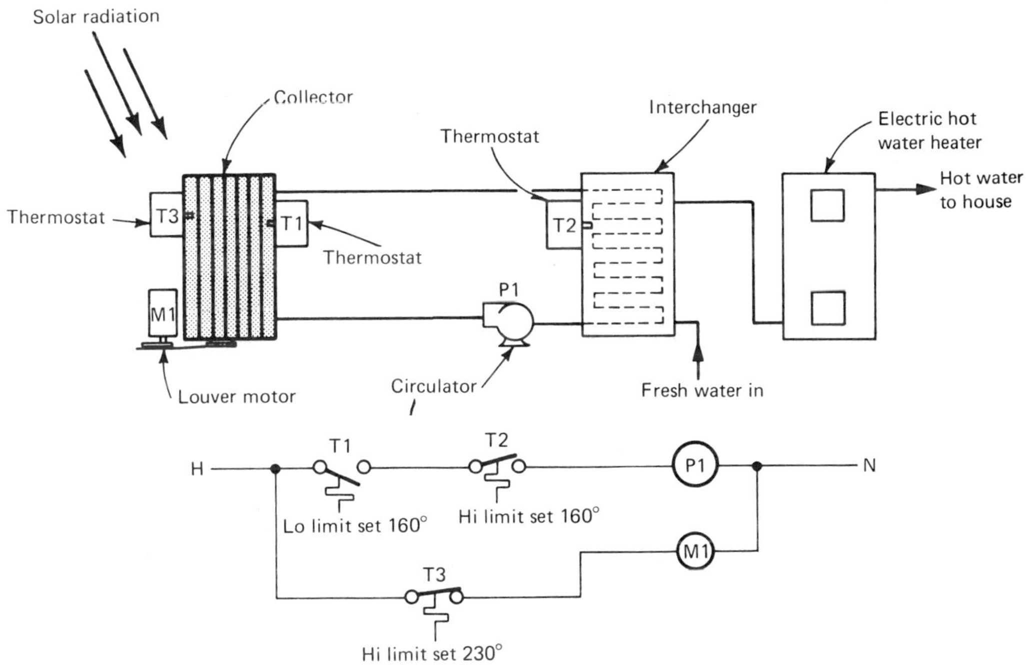

Figure 12: *Simple Series-piped Solar Energy Domestic Water Heating System*

into the home for use in an interchanger. This interchanger effects heat exchange between the heated water-glycol solution and incoming fresh water to preheat the cold water in order to lessen the need for heating within the electric water heater. The interchanger might be as large as the electric water heater itself and piped in series with it, so that fresh water would pass first through the interchanger and then through the electric hot water heater. In this manner, no heating within the electric unit would be necessary as long as water coming to it from the interchanger was as hot as the water heater's thermostats required. Or, the interchanger could be piped parallel with the water heater, with a three-way control valve used to bypass the electric unit as long as hot water was available from the interchanger. We will examine the electrical controls that might be used in a residential installation that features these two alternatives in control of solar hot water heating.

In the series-piped illustration of Figure 42, solar radiation heats the water-glycol solution standing in the collector. When this standing mixture reaches 160°, thermostat T1 closes and circulating pump P1 is started. After domestic water in the interchanger reaches a temperature of 160°, T2 opens and P1 stops. Collector high-limit thermostat T3 opens at 230° to permit the collector louvers to close, thereby shutting down collector activity when T2 is satisfied and collector temperatures rise excessively.

In this modest little system, electrical energy would still be required to maintain the hot water heater tank temperature when no hot water was being used. The water heater could be wrapped with insulation to minimize operation of electric heating elements, so that heater tank temperatures would be maintained when no hot water was being drawn. And another circulator could be installed between the interchanger and the electric hot water to draw from the interchanger's solar heat source to maintain electric water heater tank temperature. The setting of T3 would be elevated above the normal boiling point of 212° due to circuit pressurization and the addition of a glyclol solution to the interchange water circuit, both of which raise the boiling point of the solution. Depending on the water-gylcol mix ratio, T3 might even be set as high as 230°.

On a demand for hot water, such as when taking a shower, fresh water would enter the interchanger and displace the hotter water, which would flow into the electric water heater. This hotter water would pass through untreated, in terms of heating with electricity, because it was already at 160°. Even if the interchanger water was not that warm, benefits would still be derived due to its heat content. At night and on cloudy days, thermostat T1 would stop circulating pump P1 anytime that collector temperature was below 160°, to prevent reverse action that might rob the interchanger of energy stored in its warm water.

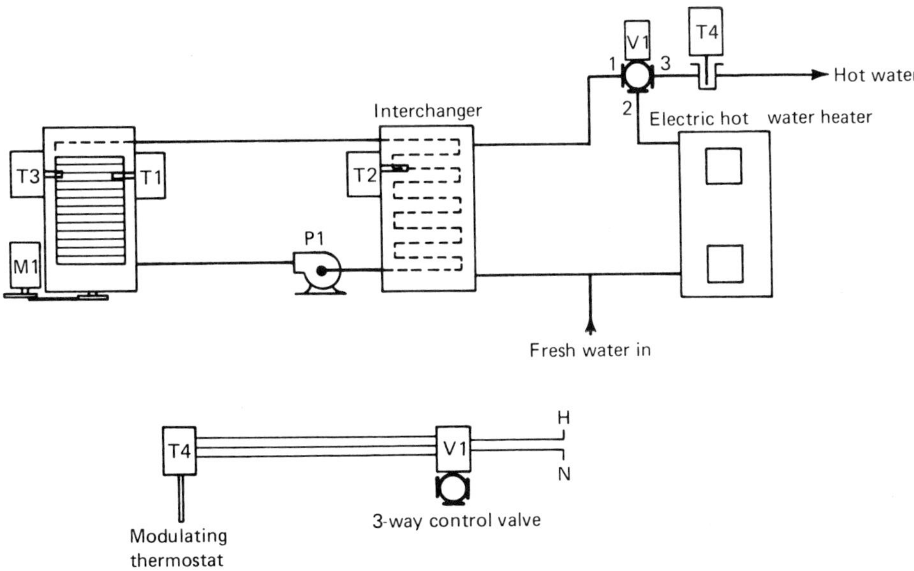

Figure 43: *Simple Parallel-piped Solar Energy Domestic Water Heating System*

An equally effective alternative to the series-piped arrangement is the parallel-piped arrangement illustrated in Figure 43, wherein interchanger and electric hot water heater each have incoming fresh water piped directly to them. Control of circulator P1 and louver actuator M1 is the same as in the series-piped arrangement, but two additional control devices are used: modulating thermostat T4 is installed in the hot water supply to the house, and 3-way control valve V1 is interposed between the two discharges of the interchanger and the electric hot water heater. V1 is modulated by T4 to mix both discharges to maintain its 140° setpoint. When T4 is satisfied, it positions V1 to admit flow from the interchanger through inlet 1, while blocking flow to inlet 2. Outlet 3 is common to both inlets 1 and 2. Both thermostat T2 and the electric water heater controls are set for 160°, so assuming equally hot reservoirs, T4 will always select hot water from the interchanger. Should interchanger temperature begin to drop below 140°, T4 will begin to modulate V1 to mix enough water from the electric hot water heater to bring the common discharge temperature back up to 140°. In the event collector activity is down for an extended period of time, T4 will modulate V1 for almost full flow of hot water from the electric hot water heater, mixing just enough water from the interchanger to maintain its 140° setpoint. In this manner, constant sampling of interchanger water temperature is assured, and collector activity will reoccur as interchanger water temperature approaches

160° again and T4 moves V1 back to admit flow exclusively from the interchanger.

These electrical control components are readily available through most heating and air conditioning parts wholesalers and would involve two occupations in making the installation. The plumbing contractor would install the thermal wells in the hot water and water/glycol piping, and in all likelihood, the electrical contractor would mount and wire the control devices themselves. The 3-way control valve body would also be piped in by the plumbing contractor and wired by the electrical contractor.

3

SERVICE EQUIPMENT

PURPOSE OF LOADCENTERS

The heart of your electrical system is the panelboard or loadcenter, often referred to as panelbox, fusebox, or breaker-panel. Its primary purpose is protection of the branch circuits. It stands as the interface between the utility's high-capacity service drop conductors and the smaller capacity branch circuit conductors within your home. Overcurrent devices, such as fuses and circuit breakers, are located ahead of each circuit to react in the event of an overcurrent by opening the circuit. Their duty is to prevent the introduction of high current onto wiring that would be loaded beyond its capacity in the event of shorts or overloads. It might be advisable to discuss, for a moment, conditions that would overload conductors if not checked.

One of the chief causes of outages, particularly in older homes, is electrical demand in excess of branch circuit capabilities. A circuit may be called upon to supply a refrigerator, hair dryer, clothes washer, toaster, and lighting, all at the same time. It is not unusual for an older home, that has not had its wiring system updated to be pressed into service by the needs of today's families. Loads such as those enumerated above may well require 30 amperes, too much for a 15-ampere circuit that was never intended to handle them in the first place. Properly installed, the two kitchen appliance circuits and separate washer circuit would have handled the refrigerator, washer, and toaster, as required by today's code, and other circuits would take care of the lighting and hair dryer. Needless to say, the 15-ampere overcurrent device would open and shut down the circuit in the face of such overload. Danger arises when occupants, attempting to restore power however they can, resort to overfusing in fuse panels that are not equipped with tamper-proof fuses. Such a move would permit current to flow at a rate which the 15-ampere wire could not handle, resulting in overheating and possibly fire. Proper protection in the loadcenter is essential for personal safety. The proper answer is new wiring and service change: a new loadcenter of adequate size must be installed, along with increased service conductors and new branch circuits.

Another condition that we are protected against by branch circuit overcurrent devices is short circuits. Sometimes people attempt to repair electrical fixtures or appliances themselves and wind up crossing wires, setting up dead shorts between hot and neutral conductors. When the appliance is tested and a short is present, current flow rises suddenly and would continue to immense proportions if not for our branch circuit protector in the loadcenter.

Still another situation that can arise is accidental grounding of a circuit's energized conductor. This happens when a fault occurs and power leaves the wiring and flows to ground external to the wiring system. Today's ground fault protector breakers and receptacles are designed to detect such a condition and interrupt power before harm to anyone occurs.

So we can easily see that were it not for the bridling effect of breakers or fuses in loadcenters, electricity in the home would be too dangerous to use.

UTILITY COMPANY'S INTERFACE

As we look outward from the loadcenter, we often see the utility's incoming conductors overhead, dropping from the power pole to an overhead attachment device. In older services, three different wires travel separately to three different insulators, where they are supported against the span of their weight before rising upward into a weatherhead. In newer services, triplex is used, and two insulated conductors are twisted around and supported by a bare conductor, which serves the dual purpose of bearing the weight of the span and acting as the grounded neutral back to the pole. After being connected to the house attachment device, they, too, turn upward into a weatherhead. The weatherhead covers the open end of the service entrance cable or conduit, preventing rain from penetrating. From there, service entrance conductors drop down the outside of the house to the meter socket 5 feet above ground level. Underground service can be observed rising up in conduit to the meter socket. From the meter, service conductors travel inward to our loadcenter.

Three wires are necessary to achieve 115/230-volt single phase service. There is a potential difference of 230 volts between the two hot wires and 115 volts between either of them and the third, neutral, wire. Lighting and receptacle loads are balance-connected so that each of the hot wires is loaded approximately the same. Their loads offset one another at the neutral point, and any difference is grounded out. If you were to energize just two 115-volt circuits fed from different legs, each with a single load of 1,000 watts, you could disconnect the neutral, and satisfactory operation would continue, since theoretically single loads of equal resistance value cancel one another at the neutral point, leaving no current to flow over the neutral service conductor. And 230-volt loads *are* connected leg-to-leg without requirements for a neutral. Thus, our 115/230-volt single-phase voltage power is brought into our loadcenter on incoming lugs, constructed so that the internal bussing they serve is alternated as it goes down through the panelboard, facilitating alternate connections between breakers as they are inserted. The bare neutral terminates on its separate neutral lug, to which all branch circuit neutral conductors are ultimately connected.

A brief look at Article 230 of the Code allows us to note some of the paramount requirements pertaining to services in residential applications.

1. A minimum vertical clearance of 10 feet is necessary for overhead service drop conductors, except in driveways, where 12 feet minimum is mandatory.

2. Grounded neutral conductors of service drops are permitted to be bare.

3. Where subject to possible physical damage, such as in driveways, near coal chutes, etc., service cable must be in conduit.

4. In instances where conduit is used, it shall be raintight and arranged to drain.

5. Connections at the weatherhead should be formed in a drip loop, with the weatherhead above.

6. A maximum of six service disconnects is allowed. These might be six switches, breakers, or fusible pullouts. But it shall require no more than

six motions of the hand to completely disconnect the wiring system on the premises from the incoming service conductors. In the panelboards selected as an exercise in Chapter 1, there was one main breaker in each panel, rated at 100 amperes and 200 amperes, respectively. These two represent the two main disconnects to isolate the premises wiring system from the utilities service system.

7. The disconnect means for the household must be located nearest the point of entrance of the service entrance conductors, either inside or outside. Many localities insist upon a separate disconnect immediately at the service entrance point when panelboards are located farther into the house, necessitating running service cable into the dwelling. This disconnect must contain overcurrent protection for the service conductors as well.

PANELBOARD DESIGN

There are basically three panelboard configurations pertaining to residential use: main breaker panel, main lugs-only panel, and split-bus panel, each differing slightly from one another. The main breaker panel incorporates one master breaker by means of which the power supply to the house can be turned off, thereby protecting the entire panelboard. A main lugs-only panel must have a master breaker field added, and a split-bus panel has one or two lower sections subfed from one or two field-installed main disconnects, located in an upper section . . . capable of holding no more than six disconnects, be they breakers or fusible pullout sections. The split-bus panelboard permits serving heavy loads directly through load-rated main circuit breakers exclusive of master-main protection, with smaller lighting and appliance branch circuits protected by separate "main lighting" breakers. These six or fewer master disconnects, in addition to being labeled as to what they serve, would also be labeled "Main Disconnect." The total capacity of the main disconnects should not exceed the rating of the panelboard itself. Effective with adoption of the 1981 code, however, split-bus panels will no longer be allowed as service equipment in new residential installations.

A bonding screw is provided by panelboard manufacturers to ground the panelboard cabinet work to the neutral bus. Where separate overcurrent and disconnect protection is provided at the outer end of the service where it enters the building, the neutral should be "floated," that is, not bonded to the cabinet work, and instead, a separate grounding lug bus installed and bonded to the cabinet work. In such a case, 4-conductor SER cable is required to provide separate neutral and ground paths to the ground bar in the entrance-point disconnect. And the grounding electrode conductor would extend from the entrance-point disconnect's grounded bus to the cold water pipe and grounding rod electrode, rather than from within the panelboard. This is to insure separation of neutral current path and fault current path, until they appear together where the grounding electrode conductor is installed. Remember, too, that the neutral from the service equipment inward must be insulated when our ground can be bare. SER cable provides for that also.

The split-bus panelboard affords a special feature that might not arise in residential work but frequently does in commercial work. Because the bussing in the panelboard is divided into two or three sections, it is possible to control all loads fed from one of these sections simultaneously by switching multiple branch circuits fed from the section. Consider a panelboard that is installed to handle outside lighting for a church. The panel is perhaps 120/208-volt three-phase, with its loads connected leg-to-neutral. We will assume that this particular split-bus panel is divided into just two sections, with the lower bussing protected by a 60-ampere breaker in the upper section. But the conductors from the protecting breaker in the upper section will be routed through a contactor that is attached by nipple to the bottom of the panelboard before it feeds the lower section of bussing. Thus, operation of the contactor via a control circuit will enable simultaneous switching

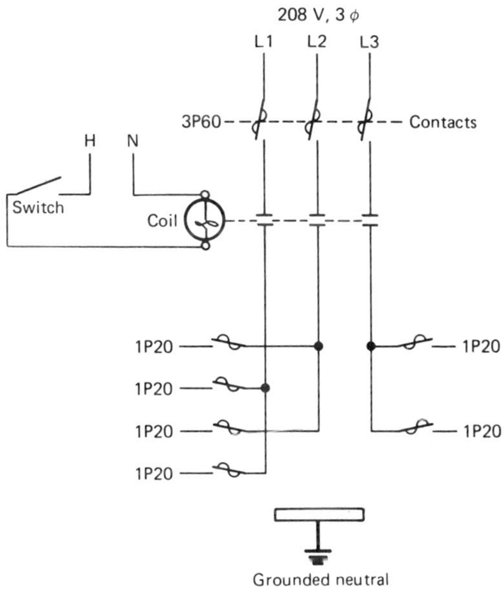

Figure 44. *Control of Slaved Section in Split-bus Lighting Panelboard*

of all branch circuits that originate from the lower portion of the panelboard. This arrangement can often be observed in church parking lot lighting, where a good number of lamps serve the church's outdoor parking facilities. Operation of a simple light switch or timing switch wired into the contactor control circuit permits control of a big lighting load with an ordinary-duty switch, with regular-branch-circuit-sized wiring extended to the outdoor lights. And each outdoor lighting circuit can be protected at its rated capacity with overcurrent protectors that are sized for the wiring and installed in the contactor-controlled lower section of the panelboard.

In Figure 44 we note the 3-pole, 60-ampere breaker first in line as we look down from overhead at the panelboard. The subfeeder from this breaker is extended on through the contacts of our lighting contactor to the separate section of the split-bus panelboard. Then, individual 120-volt circuits are routed towards the lights they ultimately connect up with. Connection of the lights is from each branch circuit conductor back to the grounded neutral in the panelboard. We can set up normal partially loaded lighting circuits and protect them at their rated 20-ampere capacity if they are wired with #12 copper wire. Breakers for our branch circuits would be left in the "on" position continuously, so that control of the lighting contactor holding coil with the pilot switch can simultaneously energize all of the parking lot lights.

PANELBOARD LOCATION

Loadcenter placement is important. Avoid locations that do not permit ready access. Indeed, your installation may be disapproved if access to the panel is not easy. Remember that in an emergency you may want to get at the panel in a hurry to de-energize one or more circuits. It must not be blocked by stored equipment or necessitate the removal of furniture or appliances. And its location should not pose a hazard in the event that inspection or repair becomes necessary. Consider a panelboard positioned in a utility area over a washbasin. Such an arrangement forces one to work over an object that is well grounded, posing a threat to the repairer's safety if an energized conductor were to be accidentally touched.

The logical place for location of the panelboard in a home with a basement is in the basement, at a readily-accessible site that is not located over any washbasins, sinks, or appliances. For homes with only utility areas, placement on a utility room wall in an uncluttered area is normal. But if the utility area is just a small furnace room, a better location might be outside in a weatherproof enclosure. High ambient temperatures, frequently encountered in poorly ventilated small areas that house furnaces, can cause overcurrent devices that protect partially loaded circuits to react, resulting in nuisance outages. A certain amount of energy is released in the form of heat when electrical current flows through conductors and devices, and warm surrounding temperatures might inhibit natural cooling and cause fuses or breakers to trip out prematurely.

Interior panelboards are either installed flush with the finished wall or surface-mounted. Widths of panels are fairly standardized for residential service, to facilitate in-wall installation between studding, spaced 16 inches on centers. Manufacturers offer front covers with 1 inch or so overlap to be used in flush installations. Flush installations are more easily accomplished, and cables stubbed into the panel cabinet, when walls are open during the rough-in stage of our electrical work. In old work, when services are changed, flush panel replacement with the addition of new circuits will require fishing of cables through attics or crawl space floor joists, and then into partitions, in order to reach the loadcenter location. Installation of surface-mounted panelboard in places such as basements can be accomplished after finished walls are up in occupied spaces, provided our cables have been previously roughed to the loadcenter location. And service changes of surface-mounted panelboards are performed somewhat more easily, owing to the "openness" of the work involved.

Cables are connected to the panelboard enclosure by use of a cable connector, discussed in Chapter 1 in the section entitled "Installing the Cables." Knockouts in the top and bottom of loadcenter enclosures are prepunched in the ½-inch, ¾-inch, and 1-inch sizes to permit removal when necessary. In addition, several "multi-punched" knockouts are offered as well, ranging up to 1½ inches in size. And one up to at least 2 inches is provided at the top and bottom, in the center, for service cable entry.

TERMINATING WIRING

Terminate service conductors first. This is accomplished more easily when there is ample free space for turning and forming. The grounding electrode conductors should come next, followed by the largest through the smallest cables. As the wiring compartment space is taken up, smaller conductors are by far more adaptable to "fitting and tucking." In breaker panels, circuit breakers will have been previously installed; their screw terminals are the termination points for energized conductors. Avoid excessive slack in the panel. This uses up room required by other conductors and hampers neat appearance. In service changes, some branch circuit conductors may not reach new termination points, so splicing is necessary within the cabinet. Use the same size wire of the same material to preserve circuit conductor sizing integrity: a pigtail of #12 wire used to lengthen a #14 conductor could mislead someone into thinking that a 20-ampere overcurrent device could be used for circuit protection, unless he were careful enough to follow the wire to the place where it joins with the old wire and detect the splice. By the same token, if #12 copper were used to lengthen a #12 aluminum conductor, the same mistaken conclusion might be drawn.

Installation of circuit breakers might be influenced to some extent by the location of old branch circuit conductors being reconnected to the new panelboard. To minimize splicing, note the entry point of the old branch circuit, its rating, and the length of its conductors. This may lead to placement of the corresponding circuit breaker on the left or right side of the panelboard, as the case may be. Deliberated placement of the breakers usually results in a neater job with fewer splices.

Dressing up the ends of old conductors being reconnected is advisable. Forming, connecting, and subsequent disconnecting during a service update frequently results in material fatigue on the bared end of the wire. It may appear mashed, excessively twisted, or even cracked. Nip away the last inch or so, and restrip insulation to expose fresh wire. The termination made will possess greater integrity, and the likelihood of overheating from conductor deterioration as a result of excessive working of the wire during previous terminations is removed.

Avoid baring any more than is necessary of the energized branch circuit conductor. Determine the amount of insulation to be removed, and be careful not to strip away too much, which would leave bare energized conductor protruding from the termination point.

Insulation-stripping tools are the best means of removing insulation from wire. A knife might produce nicks and slices on the wire, thereby reducing the cross-sectional area of the conductor. Proper stripping tools cut through the insulation but do not bite into the wire itself when the conductor is inserted into the correctly selected slot on the tool.

Routing of wiring within the panelboard cabinet should be around the perimeter, never across energized bussing. Use a splice rather than install wiring across the face of the hot bars when a wire on the left side must be terminated on the right side and is not long enough.

RELOCATION OF SERVICE EQUIPMENT

It is often necessary during a service update to relocate the panelboard from where it was originally to another point. Locations permissible in years past are no longer acceptable. It is not uncommon in older homes to find fuse boxes in closets, in small windowless furnace rooms, or in first floor or second floor hallways. In such instances, one or more of the following conditions laid down by the regulations of Article 230 of the Code may not be met: location of the service disconnect means either inside or outside of the building *nearest the point of entrance* of the service conductors: *sufficient working space* in the vicinity of the service disconnecting means; and *accessibility* to service disconnect and overcurrent devices. And in making a service change, today's Code regulations must be followed.

One choice of location that generally meets all of these conditions is outside, along with the power company's meter. Another may be in a basement immediately inside from where the meter is located. When relocation is necessary, existing branch circuits must be rerouted to the new service location. Or, they may be picked up through splices in the old fusebox, after removal of the interior and conversion to a splice cabinet. As often as not, the second method is the most feasible, and 3-conductor cable with ground can be used to good advantage to set up multi-wire circuits from the new service equipment back to the old fusebox, now gutted and converted to a splice box. Multi-wire circuits help to minimize the number of cables required between new and old service locations, and this can be important when access to the old box is difficult. Then, the neutrals of two old circuits that are using the same size wire can be connected to the single, common neutral of the 3-wire multi-wire circuit, *provided* that the two energized conductors of the 3-wire cable are connected to L-1 and L-2 in the new service equipment. Hence, two 3-wire cables with ground could be used to pick up four 2-wire cables back at the old panel location.

In homes where second-floor locations were selected for the original old service and access is prohibited because of insufficient attic clearance, it might be better to abandon the old box after gutting it, and cap off each conductor. Then, conduits can be installed on the outside to locations where a receptacle is positioned against a perimeter wall, whereupon they can penetrate the wall, enter the receptacle box, and pick up the old circuit there. Where this is done, each capped-off circuit in the old fusebox should be identified as an extension of the new circuit in the new service that feeds it. In this manner, confusion as to where the wires go and what they are for can be prevented from arising at a later date.

Labeling is mandatory as part of the service change in most localities. This insures a directory in the panelboard for future reference by owners or occupants as to which overcurrent device protects which circuit, and what each serves. Then, in the event that de-energizing a branch circuit becomes necessary for any reason, the homeowner will be able to determine which breaker to open or which fuse to remove.

Double-pole breakers that serve as main disconnects in split-bus panelboards should be labeled as main disconnects. No more than six main disconnects may be installed in any one branch circuit panelboard in any residence. When two panelboards are installed, a main breaker in each one permits isolation of each panelboard individually and completely.

Labeling should clearly denote the involvement of two breaker spaces when double-pole loads are being identified. Handle ties unite the handles of double-pole breakers for common tripping and complete isolation of both legs of the two-pole circuit whenever the double-pole breaker is opened manually or from a reaction to overcurrent. Identification on the panelboard directory should signify utilization of two-pole spaces for the 230-volt load served by the double-pole breaker. Oftentimes, numbering is embossed into panelboard interior frontworks, making reference from a similarly numbered list perhaps the most convenient method of relating the directory to the breaker in question.

Ground fault circuit interrupter overcurrent protectors should be tested periodically. The manufacturer's instructions, which are posted to the inside of the panelboard door after the GFCI device is installed, advise homeowners to push the test button routinely. Be sure to display these admonitions as the manufacturer desired.

CONVERTING AN EXISTING OUTSIDE SERVICE

A service change may have to be performed in circumstances where the previous installation made use of an outside weatherproof panelboard. If so, it may be found that the earlier work made use of cable connection to the back of the weathertight panelboard. Removal of the old panelboard will mean making provisions for reconnection of these existing circuits to the back of the new panelboard, as their elevation may be too high for any alternative entry into the new service equipment.

A more feasible solution would appear to be gutting the old weathertight box and leaving it in place for use as a splice box for picking up the old circuits without disturbing their connection to the enclosure. Installation of the new weathertight panelboard would be to one side, with a nipple connecting the two. Then, new conductors can be extended from overcurrent protective devices in the new service equipment through the raceway into the old box, where they will be jointed with the existing conductors. Multiwire circuiting might be employed here, too, to minimize the size of the nipple required between the two boxes. New circuits can be routed from the bottom of the new panelboard through EMT into the basement or crawl space. Non-metallic-sheathed cable can be used through this EMT, which would only be long enough to give us entry into the floor joists, where non-metallic cabling is normally employed, anyway. A bonding locknut may be required on the connector that secures the EMT to the new panelboard to insure grounding of this short section of EMT and conduit body. A bonding locknut provides for electrically securing a grounding conductor to the locknut, which in turn is setscrewed to the threaded end of the connector for mechanical and electrical bonding of the grounding integrity of the connector and its associated EMT.

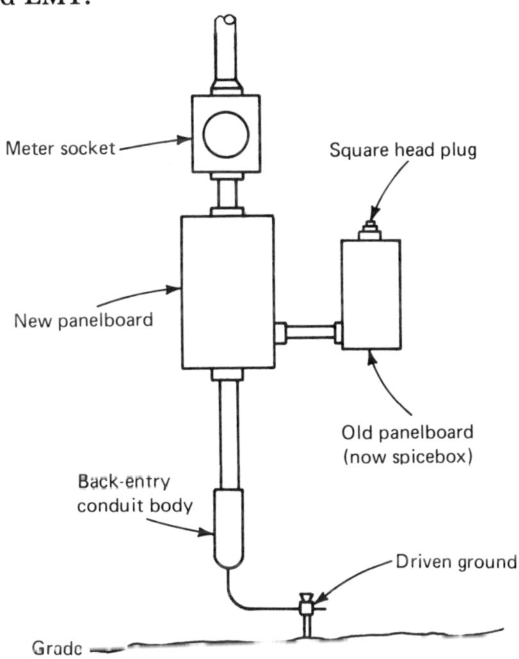

Figure 45: *Elevation of Outside Service Equipment that Reuses Old Weathertight Enclosure*

Figure 45 shows the new panelboard mounted to the left side of the old box, with the new meter in place just above. Bonding locknuts would be utilized in the bottom of the meter socket as well, and a grounding conductor would touch down on the grounded bussing of the meter socket.

The entrance where the old service cable used to come into the old panelbox must be sealed against the entry of water. Quite often it will be found that a 1-inch threaded boss was welded to the top of the old box for screwing in a weatherproof cable connector. Now that the old service cable is done away with, this opening needs to be closed. A 1-inch square-head pipe plug works very well, but it should be doped before insertion to assist in making a watertight seal with the mating threads of the boss and plug.

For the driven ground rod which is undoubtedly required in your community, consider running the grounding electrode conductor down the EMT and out the bottom of the conduit body. A 5/16-inch hole in the very bottom of the conduit body will allow exit of the grounding electrode conductor, and the short run to the driven electrode will be all the less noticeable.

A short section of nipple of EMT should be used from the conduit body to penetrate the outside wall for protection of the non-metallic-sheathed cables at the place where they enter the house's crawl space or basement. Concrete can then be used to patch around this short section of raceway where it goes through the house. This should hold true for most situations, but a review of your intentions with your local electrical inspector is in order to be sure of your compliance with regulations in the work undertaken in his jurisdiction.

RACEWAY SYSTEMS

Occasionally, raceway systems become necessary in areas such as basements or utility rooms where wiring must be installed to serve outlets after block wall construction is finished. When finished ceilings are also involved, a raceway system for the conductors is perhaps our most feasible alternative. This raceway or conduit system would be installed so as to be exposed on the surface of walls and ceilings. Classic cases that necessitate the use of a conduit system are installation of receptacles around the finished basement room where painted block walls and drywall ceilings form the interior finish scheme, and situations where a service change is to be made from an existing fusebox whose branch circuit runs rise straight up into the concealed ceiling area. We shall consider the latter in order to develop a scenario that mandates the use of an exposed raceway system, so that we can discuss the techniques applicable to installation of thinwall conduit, otherwise known as electrical metallic tubing (EMT).

The decision to finish off the basement of this older home was arrived at, say, three or four years ago by a previous owner whose electrical needs were being met satisfactorily by the existing service equipment. Consequently, no concern was focused on the presence of the fusebox surface-mounted on the wall. It was small and not very obtrusive, so it was easy to ignore its presence. Besides, its remaining exposed in the finished scheme was necessary to permit access for service, and the decision to leave it alone was made. Ceiling material was installed, subsequent to which only 2 feet worth of six armored cables could be seen rising from the fusebox top into the concealed spaces of the ceiling. One cable was the service entrance cable, one was the electric range cable, and the other four were non-grounding type, 2-wire general lighting circuits. Hot water heating for domestic needs was through a summer-winter hookup with the oil-fired, hot water heating system, and no air conditioning or dryer was involved. Now the home has been sold, and the new owner needs a window unit air conditioner outlet in the second-floor hallway and a dryer outlet, and he desires an electric water heater. Upon contacting an electrical contractor, he discovers that not only is he a candidate for a service change, but, additionally, a kitchen circuit to serve small appliances and a separate washer circuit must be installed to comply with minimum residential electrical standards that apply whenever a rewire is performed. The term 'rewire' is frequently used to denote a change in service equipment and necessary in-house wiring to bring the home in line with minimum property standards set by the local municipality that has jurisdiction. Such standards are enforced when residential housepower is increased at the time of a service change or rewire. For this work to be performed in the setting we have described, exposed conduit will have to be run on the walls and ceiling of the basement to support the electrical improvements being made. It's either that, or open the ceiling for the installation of concealed wiring.

To me, a conduit system originating from a surface-mounted panelboard is an attractive assembly, because when properly done, it reflects the craftsmanship of the installer, with its 90° bends and offsets. But one would probably have to be an electrician, like myself, to appreciate the beauty in it. Nonetheless, the panel and conduits can be painted the same color as the finished wall, and this does tend to disguise the surface-mounted electrical work and help to blend it in so its prominence is diminished.

In examining the work to be done, we see the existing service entrance conductors entering the attic, high above at the point where connection is made with the utility's service drop off the pole. Then they drop, concealed in the wall to where they emerge just above the meter socket on the side of the house. Then, from the bottom of the meter socket, they duck back into the structure and run concealed to the old fusebox location on the front basement wall. The last two feet of service cable is visible in our group of cables just above the fusebox, dropping out of the finished ceiling. In the municipality where we are to perform the work, all service conductors must be outside the home, with only through-the-wall runs permitted straight into the service equipment, unless the service entrance first terminates at an overcurrent device at the point where the service conductors enter the building. This is good logic, because the available short circuit current straight off the utility company's overhead pole-mounted transformer is introduced into the residence when unprotected service entrance cable is run within the home. 'Unprotected' in this sense means 'unfused' as to the amount of current that can rush in when a short circuit occurs. So it turns out that the existing entrance conductors are in violation of today's regulations, due to their presence within the interior spaces of the home itself. Accidental conductor damage to a branch circuit cable that resulted in a short circuit would merely activate the branch circuit overcurrent protective device, be it fuse or breaker. But the unbridled service entrance conductor may have short circuit current potential of 10,000 amperes, which would create tremendous thermal and magnetic forces that would result in short circuit to ground in the event of conductor damage. This could be extremely destructive, so protecting service conductors before they enter the home, at a level of 200 amperes in the case of 200-ampere service equipment, limits the destructive forces by assuring reaction of a protective device in the event of accidental grounding.

The new service entrance cable will drop from overhead outside the home to the new meter socket and be secured to the outside wall. At this point, we could elect to set a 200-ampere fused disconnect in a weathertight enclosure, nippled to the meter socket, before extending the protected service conductors on to the existing fusebox location where we would place the new panel. But we decide, instead, to set the new panel back-to-back with the meter location, where we can drop off the bottom of the meter socket and then penetrate the wall and enter the back of the new panel. In this configuration, a sufficiently large conduit must be run from the new panel to the old fusebox to pick up the old circuits. We will gut the old fusebox of bussing and fuses and convert it into a splice box by making joints to pick up the existing circuits. Saving the cost of the fused disconnect outside can help hold down job costs. Once the service change work is completed, reconnection to the existing service drop from the pole is allowed until inspection can be made by the locality's electrical inspection personnel, after which the utility must replace the overhead service drop.

Four 15-ampere circuits and one 50-ampere range circuit originated from the old fusebox, so two 15-ampere multi-wire circuits of six #12 copper conductors are used to serve the four old general lighting circuits, and three #4 copper conductors are selected to pick up the old range circuit through the interconnecting panel-to-panel raceway. Derating of conductors is required per the notes to tables in Article 310. A factor of 70% must be applied to 9 conductors in a common raceway. Conduit sizing is based on selection of wire with THHN insulation, 40% fill, and tables in Chapter 9 of the National Electric Code. Table 5 shows us that six #12 and three #4 copper conductors with THHN insulation constitute .32 square inches, and Table 4 reveals a need for 1-inch conduit in the 40% fill column. If the municipality were to require a bonding conductor between the old, gutted fusebox and the new panel, conduit size would be increased to 1¼ inch, but generally the conduit itself with its approved connectors and couplings would be the acceptable bonding path. Additionally, the old fusebox grounding electrode conductor would be removed and the old box bonded instead to the new box casing to insure grounding of the converted splice box. So our first conduit run would be with 1-inch conduit from panel to box. Bends must be made with an approved bending tool, so that the internal diameter of the raceway is not effectively reduced, and the radius of the inner edge of the field bend must meet the requirements of Code Article 348. We note further that the Article requires that supports be affixed every ten feet, and within three feet of every box, cabinet, or fitting. No more than the equivalent of four 90° bends may be made outlet to outlet, outlet to fitting, or fitting to fitting. The raceway must be installed as a complete conduit system before the installation of conductors, and splices may be made only within boxes or conduit bodies, never within the EMT itself. Conduit bodies are defined in Article 100 as a separate portion of a conduit or tubing system that provides access through a removable cover to the interior of the system at a junction of two or more sections of the system or at a terminal point of the system.* We elect to make a horizontal run with this initial piece of EMT and turn one inside corner that necessitates one 90° bend. At the panel and old fusebox, two offsets are formed to allow offsetting the EMT from the wall enough to make connection to the metal cabinets through 1-inch knockouts. Our EMT is securely fastened to the wall with 1-inch straps.

*Reprinted by permission from NFPA 70-1981, National Electric Code®, copyright © 1980, National Fire Protection Association, Quincy, Mass.

Runs of EMT will also be required to serve the dryer and water heater if they are located in the basement. The washer circuit and dryer circuit might share a common raceway, and if the water heater is adjacent to the laundry equipment, it might also have its circuit extended into the general area in the same conduit run. For the kitchen circuit, holes will have to be cut in the basement ceiling directly below the partition to be chased for installation of a receptacle outlet box. Non-metallic cable can be extended down from the receptacle outlet box into a basement ceiling-mounted 4-inch junction box with blank cover. Where a cable-to-EMT conversion is made, a cable connector must secure the cable to the junction box. The second-floor air conditioner receptacle outlet will be piped in solid, that is, in a continuous raceway system from air conditioner receptacle outlet box to the service panel. We elect to install a surface-mounted receptacle box and penetrate the wall with a steel, threaded nipple of the appropriate length to permit locknutting to a ½-inch knockout in the box backplate. Then a weatherproof conduit body outside turns downward after being tightly threaded to the through-the-wall nipple. EMT drops outside to another conduit body, and wall penetration again takes us into the panel. Thus, the raceway system is mechanically and electrically continuous from receptacle outlet box on the second floor all the way down the outside wall and into the service panel, setting up good mechanical and electrical continuity along the complete path.

In such manner, we have visualized residential electrical work that involved conduit and the circumstances which led to the necessity for its use. Conduit offers an opportunity for the electrician to demonstrate his craftsmanship, as a good job reflects proudly upon the installer whose talents were brought to bear upon the work. In an installation of such quality, it will be noted that each vertical and horizontal run will be plumb or level, as the case may be. And all aspects of the job will exhibit thoroughness, neatness, and completeness.

CONNECTING TO OVERHEAD SERVICE DROPS (WORK RESERVED FOR QUALIFIED PERSONNEL ONLY)

Different parts of the country may find utilities differing in their policy toward service entrance cable from the weatherhead down. When rewired residences must be tied back in to overhead service drops, the weatherhead and its installation may sometimes become the responsibility of the electrical contractor, as does the service entrance cable down to the meter socket. A few special precautions are worth noting that pertain to disconnecting and reconnecting service entrance cable to the utility's service drop off the pole.

Call the Utility to Break the Meter Socket Seal

Diversion of power has become a problem in many areas, as the utility suffers the loss at the hands of people who attempt to bypass the electric meter. Because of this, breaking of meter socket seals for any reason is usually reserved to be performed by the utility's own personnel. In this manner, they can log and record meter seals that have to be temporarily removed for some reason or other. The utility is glad to cooperate, and it will usually schedule a representative to break the seal whenever you require, if you provide sufficient prior notice.

Pull the Meter Before Cutting Overhead Leads

Removal of the meter from its socket positively isolates all load from the service drop so that cutting of the service entrance leads will produce no arc. When individual conductors are cut, no arc will be formed if no current is flowing. Creation of an arc occurs when wattage is being consumed, and severing the wire breaks the circuit. As long as the meter is pulled, no power is flowing at all, so no arc will be produced.

Use a Ladder Constructed of a Nonconductive Material

Inadvertent touching of one of the service entrance leads will not cause a circuit through you to earth if the ladder you are on is nonconductive. Selection of a ladder made out of wood or fiberglass is a good choice from a safety standpoint. When you are on the ladder, there is no conductive path between you and earth, as there would be on a metal ladder. Your body cannot be part of a path to earth when the ladder will not form part of the circuit also.

Reserve the Performance of This Work for Dry Weather

An elective service change can be performed in dry weather. There is no need to attempt such work in wet weather, when danger exists from naturally conductive situations that occur then. Wet surfaces, skin, and clothing may present hazardous situations due to their potentially conductive nature from being wet. Postpone any work of this nature until the weather is clear and dry again.

Wear the Proper Protective Clothing

At an overhead service entrance, there is a potential to ground of 115 volts and a potential between energized conductors of 230 volts. This is not high voltage by any means when compared with pole-top transmission voltages, but basic precautions in the utilization of the proper protective clothing are nonetheless in order. Wear lineman's leather gloves with rubber inserts, and put on a Plexiglas or plastic resin material hard hat. These items hold voltage away from contact with the body should a conductor accidentally come in contact with your hands or head. They increase the odds in your favor in your potentially dangerous environment at the top of the ladder near the service drop.

Mount the New Weatherhead Above the Service Drop

Placement of the weatherhead above the point where the service drop is anchored to the house structure will aid in the creation of a drip loop to forestall the entrance of rainwater into the weatherhead and down the service cable jacket. Keeping the weatherhead above the service drop assures that no water will migrate onto the cable to reach the point where the jacket of the service cable would be stripped away, exposing conductors intended for termination to the service drop conductors.

Leave 36 Inches of Slack Protruding From the Weatherhead.

This minimum prescribed length of service cable conductor insures that the utility will have no difficulty in terminating the new service drop to the new service cable. Our temporary terminations may make use of split-bolt connectors, but the utility will in all likelihood use a crimping tool and crimping sleeve to make a permanent bond between the two conductors to be united. Leaving the minimum prescribed length of 36 inches provides the necessary length to permit removal of the fatigued wire created by our temporary connection, so they can start with a fresh end of each conductor. Tape the temporary joints thoroughly to insure that no short occurs should they touch in a high wind. Our connectors can be retrieved and used again if the utility is instructed to leave them behind with the owner so that the electrical contractor can pick them up next time he's out that way.

4

REMODELING AND IMPROVING

ADDING OUTLETS

Working in an existing home is far more challenging than working in a new home with respect to figuring out how to get boxes installed in walls, since the walls are already closed in. Basements, crawl spaces, and attics offer access to tops or bottoms of partitions in the occupied spaces; and drilling into the top plate or bottom plate will usually permit fishing in the cable to an outlet point.

Switch and receptacle boxes similar to those discussed at the beginning of Chapter 2 are usually employed for old work, but are provided with "ears" that allow the box, after being fitted with the cable and inserted into the hole, to be mounted against the finished surface of the wall with its ears pulled down flush. Then, special modernization clamps are utilized to prevent the boxes from being pulled out.

In one of the popular setups shown in Figure 46, a beveled-corner switch box has its ears at the top and bottom for use in modernization work.

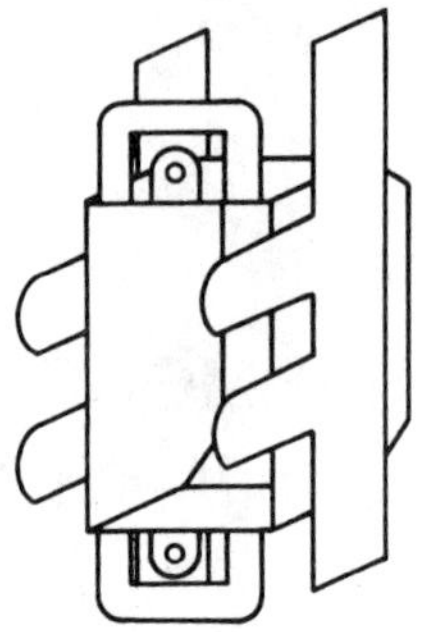

Figure 46: *Modernization Box with F-brackets*

The ears are removable, so that the box can also be used in new work. The F-brackets, shown at each side, are purchased separately and are designed to be installed beside the box in the wall cavity. Then, the four protruding tabs are folded over into the box, preventing it from coming forward. The brackets are longer than the opening that is cut for the box and therefore catch the wall surface from the inside, while the box ears catch the wall surface from the outside. The necessity for mounting against a stud is eliminated, as this method is designed specifically for mounting in a cavity.

When cutting the hole for the box, be sure to make allowance for the receptacle or switch mounting screw, which utilizes the threaded hole inside the ears. This will assure that the box pulls down flush on the wall and prevents the mounting strap screw from encountering wall material. The box needs to be flush also, to prevent the trim plate from standing off from the wall.

In another popular setup, illustrated in Figure 47, the bracket is furnished with the box. This time, it's a U-shaped bracket drawn up behind the box with a screw, accessible from inside the box, to grip the wall from behind, while its ears flatten out against the wall, as in the other arrangement. The U-bracket is compressed to the sides of the box upon insertion into the cutout for the box, and once pushed through, it returns to its original shape, rendering it larger than the opening.

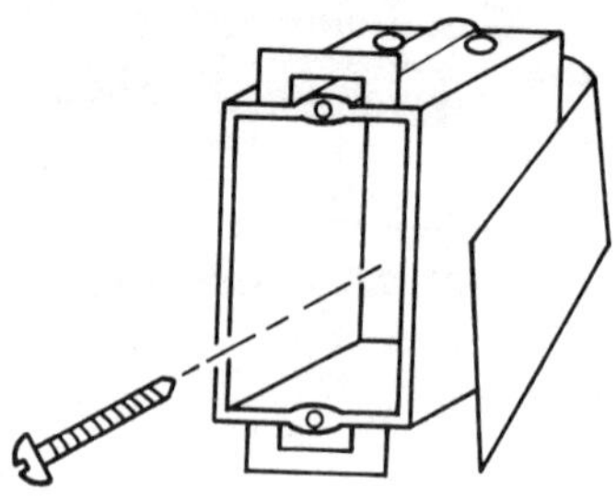

Figure 47: *Modernization Box with U-bracket*

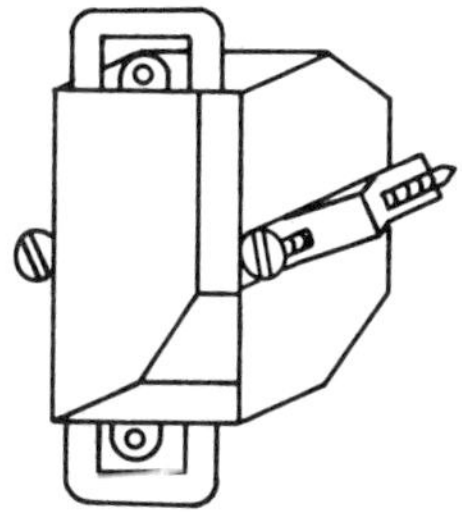

Figure 48: *Modernization Box With Wedge-wings*

A third popular arrangement, shown in Figure 48, features wedge-wings on each side of the box, drawn up by a screw and turned from the front. These track-guided wings expand as the screw is turned and snug the box flush against the wall as they exert pressure on the back side of the wall during tightening. Tightening is generally best accomplished by alternately turning one screw and then the other, bringing both wings out together to exert about equal pressure on both sides of the wall from behind, thereby insuring leveling of the box.

Since adding receptacle or switch outlets usually makes up a significant portion of a residential rewiring job, these boxes represent an efficient method of in-wall installation in older homes. Cable clamps are available in these boxes as well, and are necessary in old work, where stapling of the cable within the wall is frequently impossible without opening the wall. In all cases, the cable must be secured within the box, to prevent stress of support from being imparted to conductors and device termination points.

The first step in installing additional outlets is the selection of locations in parts of finished walls that are clear of studs. A stud-finder might be put to good use here, or small pilot holes can be drilled along a horizontal plane to find open wall space for installation of our "old work" boxes. Small pilot holes are easily filled and patched. Once the wall location is decided upon, check for access to that particular point from above or below, depending on whether attic-access or crawl space- or basement-access is to be attempted. You should be able to locate the two-by-four that forms either the top or bottom plate of the partition to be entered.

Take careful measurements in the room above from the intended outlet location to an easily recognizable location marker, such as a duct going up into the room or a radiator pipe, both of which can be observed in the finished room and in the access space. Then, using the same measurements, plot the point to drill into the partition. Drill a ¾-inch hole into the partition void. If you are working from above, drop a string line with a small weight attached through the hole until it bottoms out, and tie it off. If you are working from below, insert a folding rule up into the partition to the height of the intended outlet. The rule should have a 3/16-inch hole drilled 3/4-inch from the tip, into which a pull wire can be inserted to facilitate pulling back the cable.

Back in the finished room, hold the outlet box front-side to the wall in which you intend to install it and, using a pencil, draw its outline on the wall, properly leveled. (A torpedo level with magnetic insert will self-adhere to a metal box, making leveling easy.) Then, using a keyhole saw or electric sabre saw, cut out the outline of the box, after first drilling four 1/4-inch starter holes for the saw blade, one at each corner. When the cutout is removed from the wall, you should see your stringline or folding rule, ready for the cable or pull wire to be attached for pulling back.

Sometimes, horizontal bracing is encountered in partitions, and it blocks the path from partition penetration point to outlet location. When this is the case, open the wall across the face of the two-by-four brace, making a thumb-sized incision through the finished wall material. Once the wooden member's edge is exposed, use a 5/8-inch wood speed bit to notch the brace enough to permit recessing the cable into it in such manner that the cable is flush with the edge of the brace. Then, fish the cable to one side of your incision. Pull it out into the room far enough to reach the outlet location, plus 6 inches for conductor slack in the box. Then, insert the cable across the face of the wooden member into the other side of the incision, and fish it on out into the opening intended for the outlet box. Staple the cable into the recessed area, and cover it with a nailstop, a small, flat piece of 1/16-inch thick metal installed to bridge the recessed area of the brace. The incision may later be patched and painted.

Figure 49 shows our cable in the void between two studs, together with the horizontal brace to be crossed. The incision will probably be no larger

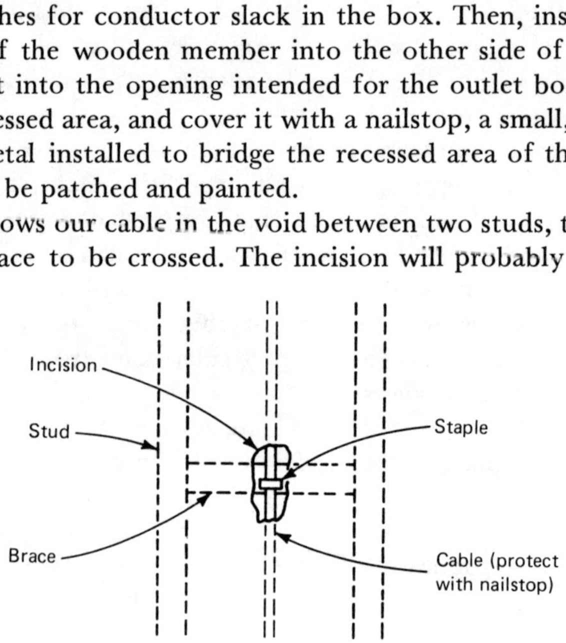

Figure 49: *Crossing Concealed Wall Brace*

than 3/4 inches by 3 inches, and it can be professionally patched in a two-step patching operation. The nailstop is not shown, but it should be used to prevent penetration of the cable by nails that might be driven into the wall in the future to hang pictures or install chair rails.

When outlet locations cannot be fished from above or below, concealed cross-wall cable installation can be utilized to run cable from one existing outlet to another. This requires opening the wall across the face of each stud in the path of our route. The same thumb-sized incision used for crossing horizontal bracing can be made to cross each vertical stud, and the cable can be fished from opening to opening until we reach the new outlet point. Patching plaster can really be used quite easily and professionally with patience and a little care to restore the finished plaster or gyp board to its original state. Patching of wood-paneled walls may require wood filling and staining and may not be altogether satisfactory. But openings could be cut at chair rail height, with chair rail installed later to hide them.

Should the intended outlet require a special circuit, such as for a window air conditioner, and not be accessible from above or below, consider running conduit outside for placing the receptacle as desired.

All in all, patience coupled with imagination and applied to the installation of additional outlets can result in a neat and effective job. Even if it is not desirable to open walls, additional convenience receptacles can be introduced into a room through the use of surface raceway, installed on the surface of the wall or baseboard. Surface raceway is flat on the back side so as to hug the wall and is available with manufactured inside ells, outside ells, and right-angle ells, to permit turning corners and making right-angle turns. Surface-mounted outlet boxes are available also, not only to permit surface-mounting of switches, receptacles, and fixtures, but also to permit conversion of a flush outlet to a surface outlet so the raceway system may be extended from it.

Surface raceway systems are marketed in standard 10 foot lengths, just as in conduit, and they can be cut to whatever lengths are necessary and installed along the top of the baseboard in many rooms so that they are barely noticeable. Extreme flexibility is afforded through the use of surface raceway, for outlets may be positioned virtually anywhere. Raceway can be painted also, to blend in with whatever wall surface it lies on.

Individual conductors are pulled into the completed raceway system after it is in place. Snap covers are left off of joints in the raceway system to permit insertion of a fish tape and subsequent pulling in of the circuit conductors. A green or bare grounding conductor must be pulled in also, to preserve the grounding integrity of the wiring system throughout the house.

Mounting clips for surface raceway systems offer holes for anchorage to wall surfaces with screws. Wood screws can be used either directly on wood-finished walls, or with plastic expansion shields on plaster or gyp board walls, or with plastic or lead expansion shields on masonry walls. Toggle bolts serve well, too, where there is a wall cavity to permit opening of the hinged wing nut.

ATTIC LIGHT

How many times have you gone into the attic with a flashlight to get something? An attic light, controlled from below, is very nice, and provides a great service for the relatively small cost involved. Many existing homes do not feature a light in the attic, but it is fairly easy to add in most instances.

To be most serviceable, the switch should be located in the vicinity of the access to the attic, be it permanent stairs, pull-down disappearing stairway, or scuttle hole. Because the light in the attic is not visible with the access closed, the switch should incorporate an indication of when it is in the "on" position. This would alert occupants passing by that the remotely located attic light was indexed to be "on." This pilot indication is particularly helpful when the switch for the attic light is ganged with other switches that control other lighting. With several to choose from, it is all too easy to switch on the wrong one inadvertently. Without any indication that a light had come on or was supposed to be burning somewhere, the switch would in all likelihood be left in the "on" position, as the unaware person immediately addressed himself to the next switch in an attempt to turn on light in the area in which he was interested.

In actuality, the pilot light type of switch turns on two lights, the one in the attic, and the small pilot lamp incorporated into the switch itself. This necessitates the presence of a neutral within the switch box to permit the pilot lamp to burn when the switch is in the "on" position. The circuit is simultaneously completed to the light bulb in the attic fixture and the small pilot lamp within the switch.

In Figure 50, the switch shown features three screw terminals. The one labeled "n" will in all probability be silver or essentially white in color, establishing it as the termination point for the grounded conductor, or neutral. Constant hot is terminated on the gold terminal, labeled "h," and the switched leg to the attic light is terminated on the black screw terminal,

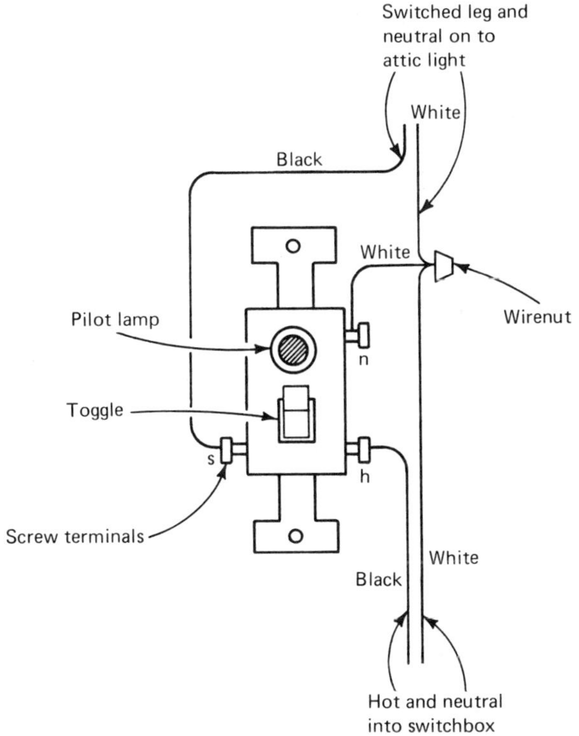

Figure 50: *Light Switch with Pilot Indicator for Remote Control*

labeled "s." Manufacturers' markings for the various terminals may vary slightly, except for the identified terminal upon which the neutral is terminated, but each one will be sufficiently marked to permit proper terminations.

An interesting note concerning this particular type of light switch becomes apparent when we examine how the switch operates and what the associated internal wiring schematic that invariably accompanies the switch looks like. Because the circuit is completed through the pilot lamp and back to neutral, our pilot light does not actually tell us that the remotely controlled lamp is on, only that it should be. Or more correctly, it tells us that our switch is in the "on" position. For even if the light bulb in the remote attic fixture were removed, the pilot lamp would still burn.

Another alternative for us to exercise concerning indication of the status of the remote lamp presents itself in the form of a standard two-terminal type of light switch that features a glowing toggle whenever the switch is in the "off" position. The usual purpose of this type of switch is to facilitate locating it in a darkened room, as it is readily observed by its illuminated toggle. This is the same idea as in doorbell pushes that glow rendering them visible from the outside at night to make it easy to find the button to ring the doorbell. A tiny neon indicator is built into the toggle handle and internally wired in parallel with the two terminals on the switch. With this arrangement, when the contact within the light switch is opened to extinguish the controlled lamp, the highly resistive small pilot lamp is placed in series with the switched leg through the lamp, and the filament within the bulb merely acts as a conductor for the tiny amount of trickle current required to cause the miniscule lamp within the toggle to glow.

We can take advantage of this kind of switch to give us our indication in reverse: the remote lamp is on when the pilot is out. This method does not require a neutral inside the switch box; our remote lamp could be fed hot, with a mere switch leg dropped to the switch box. A by-product of this application is indication of whether the bulb in the remote lamp holder is good or not. For the brilliance observable within the toggle is dependent upon the filament inside the remote lamp being intact. When the bulb burns out, the series circuit for our tiny neon indicator no longer exists through the bulb.

In considering our intended attic light installation, let us assume the handy location of a receptacle in the hallway below the attic opening, but no other switches in the immediate vicinity of the opening. Our preference is for the light switch to be right at the attic opening, perhaps just above the hall receptacle. We will tap into the second-floor circuit within that receptacle and extend it through our switch on into the attic to serve the new attic light that we are putting in (see Figure 51).

The home is of recent construction. The walls are drywall and will be fairly easy to cut for installation of the required switch box. The attic is not floored, but there is a great deal of insulation between the joists. We will need to locate the top plate of the partition in which we plan to place the switch box. Our decision is to position the switch immediately above the hall receptacle and take advantage of a straight rise to the attic above without having to concern ourselves with crossing wall studs.

Take a careful measurement of the distance from the closest edge of the trap door to the receptacle. We will come up with a right-angle measurement that will allow us to pinpoint the spot in the attic immediately above the partition and, consequently, the receptacle's location in it. Bore a ¾-inch hole down through the top plate, which will be observed when we pull back the insulation. A long-sleeved shirt is advisable, as it keeps the insulation off your arms. We will install our switch box along the vertical line from where we drilled the hole to where the receptacle is located. We then go back to the hallway below to cut the wall for the switch box.

Mark the opening to be cut for the switch box directly above the existing receptacle. This should keep us clear of studding, although we will expect to see a stud right at one side of our new opening. The 52-inch height should spot our hole just above horizontal wall bracing.

After disconnecting power, fish a piece of cable between the new opening and the existing receptacle box. Most likely, the circuit into the receptacle will be 15 amperes, so continue it on to the new switch with the same size wire as was brought into the receptacle box. 15-ampere circuitry permits us, of course, to use #14 cable. Next, using an electrician's fishtape, pull new cable from a self-feeding box up through our new opening into the attic, and run it back toward the eaves alongside the joist. Do not rise straight up to the overhead attic fixture outlet box; cable must flow with the structural members and be properly supported. Go under ducts that cross attic joists, and come back along the rafter line to the ridgepole location of the octagonal box for the attic porcelain fixture.

Another consideration as we think in terms of the switched leg into the attic is whether any future plans might call for the presence of a constant hot and neutral for later finishing of portions of the attic or for a future attic exhaust fan. Since we are running new cable anyway, I would suggest using 14/3 with ground. In that case, the red insulated conductor becomes the switched leg, and the black insulated conductor is capped, after being made up hot in the new switch box below. This will leave a constant hot source within the octagonal box outlet, should it be desired later on.

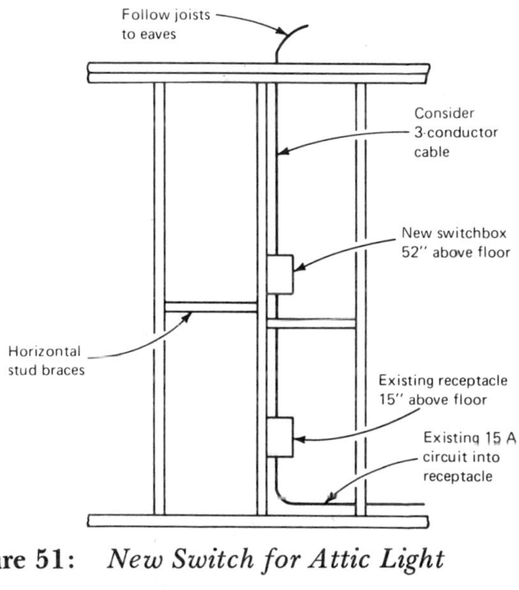

Figure 51: *New Switch for Attic Light*

Looking back for a moment, fishing of the connecting cable between the new switch box and the existing receptacle outlet would probably cause us to encounter a horizontal brace immediately below the switch box location. A ¾- by 18-inch long bit in a drill motor would permit us to drill through the brace by inserting the drill bit down through the opening for the new switch box. This would preclude the necessity to open the wall in order to clear the brace. The long length of the bit is necessary so that the minimum angle at which we want to drill does not cause our drill motor to be up against the wall, precluding the almost straight shot we want to make down through the brace.

Assuming the use of 14/3 cable for preservation of a constant hot within the attic, our connections at the pilot light switch would be as shown in Figure 50, except that the hot into the attic would be connected into a joint with the incoming hot by a Wire Nut®*, with a pigtail terminated onto the light switch terminal. Grounds were omitted for clarity, but they would be bonded together in the existing receptacle box and new switch box. In the attic, a ground would be bonded to the outlet box if the box is of metal.

ATTIC FANS

Reducing the temperature of your attic in the summertime can contribute to the comfort in your home and save you some money in air conditioning energy costs. Fans that are designed for the purpose can be connected to operate above a predetermined setpoint attic temperature, inducing outside air to enter the attic for ventilation by exhausting air out one side through gable louvers or out through the roof through weathercaps.

To be effective, the fan should be capable of moving 1,000 cubic feet of air per minute per 600 square feet of attic floor area. Usually, fans are available for these capacity requirements in 115-volt models, permitting tie-in to branch circuits already in the attic. If a pull-chain porcelain fixture is present in the attic, the fan might very well be tied in there. Cable should pass through roof rafters when run at an angle to them and terminate at the fan motor wiring compartment if the fan features a built-in thermostat. Such thermostats usually energize the fan on a rise in temperature to between 115 and 120° and de-energize the fan on a fall in temperature to between 95 and 100°.

*Wire Nut is the registered trademark of Ideal Industries, Inc.

If a remote thermostat is desired, the built-in one can be bypassed or removed, and a switch box can be installed in a preferred location in the attic, onto which a "cooling" thermostat can be mounted. Its temperature range should be from about 75 to 125°, so that the setpoint represents the approximate midpoint of the thermostat's range.

For gable installations, direction of the prevailing winds should be taken into consideration. The fan exhausts air from the attic space by pulling air in from the other side. Arranging the fan to discharge in the same general direction that compliments normal air movement patterns in your area takes advantage of prevalent wind direction to aid in ventilation of the attic space. Facing the gable-mounted fan into the wind imposes resistance on the fan when outside air movement is significant. Through-the-roof weathercap models are best located on the leeward side of pitched roofs, for the same reason.

For electrical tie-ins within the attic to procelain light fixtures, consider whether the attic light is switched from below. If so, attic fan operation will be dependent upon that switch being left on. This would necessitate either loosening the bulb to prevent the attic light from being on all the time or installing a pull-chain fixture, which can be turned off while the fan is permitted to run.

Ideally, a separate switch should be installed in the hallway below, where an existing hallway switch controls hall lighting. This circuit could be extended through the new switch to the thermostat and attic fan above, permitting shutdown from below when it is desirable to prevent fan operation for any given period of time. When not controlled from below, a switch should be installed close to the fan to permit manual disconnect.

Normal ratings for motors of the class under consideration are 115 volts, around 1.5 amperes approximately 1,500 rpm, and 1/20-hp. The Code in Article 430, Part C, addresses itself to running overcurrent and overload protection for small motors. It is generally interpreted to permit connection of 1/20-hp and smaller motors to existing branch circuits without the need for additional overload or short circuit protection if the motor is marked "impedance protection," indicating that the motor is designed so that dangerous overheating does not occur should it fail to start when energized. This is often the case with many shaded-pole induction motors in which sufficiently high winding impedance will preclude excessively high overheating temperatures when the rotor stalls in these small motors.

Motors of over 1/20-hp for larger fans will require motor protection in accordance with Article 430.

In normal summertime operation, the attic fan may start automatically around 10 or 11 A.M. and run until midnight or so, depending on the amount of shade around the home. Thermostatic control cycles the fan off when the attic temperature no longer represents a significant detriment to the maintenance of cooler temperatures in the conditioned spaces below. For air conditioned homes, this results in dollars saved, as the air conditioning equipment operates less time, since the heat otherwise radiated from warm ceilings is removed by the more economical attic fan. Additionally, items stored in the attic are less likely to be distorted or damaged from high attic temperature when attic ventilation is provided.

BATH HEAT LAMPS

Need a touch of heat in the bath after showering? Consider installation of heating lamps to take the chill off the bath area in the morning. These localized heating units mounted in the ceiling warm the area and make stepping out of the shower more pleasant.

Perhaps the most popular types of bath heating lamps are the one-bulb and two-bulb moldels. If your bath is particularly small, the one-bulb unit might be sufficient; if your bath is medium to large, the two-bulb unit would probably prove the better bet. Most two-bulb units offer two-level control, making it possible to turn on just one or both of the heating lamps, as you desire.

Let us look, now, at the installation of a typical two-bulb heating unit in the ceiling of a second-floor bath. Careful measurements are in order prior to cutting the opening in the bath ceiling for installation of the housing. Find a reference point common to both the attic and occupied level below, such as an air conditioning diffuser in an adjacent bedroom ceiling. Take a right-angle measurement to determine the desired centerline location of the heating unit. Then go to the attic and plot these measurements above the area below to see where the intended centerline falls with relation to joists. You will find that a location in between two joists must be settled for, representing the point closest to the exact center of the bath ceiling below.

Brush away or pull back insulation, and position the housing down between the joists as it will eventually appear when roughed in and mounted. Now make two very small pilot holes on the ceiling below at diagonal corners of the housing, and go back downstairs with the unit in hand. Now hold the unit up against the ceiling in between the two diagonal pilot holes, corresponding to its position between the joists. Scribe the outline of the top portion of the housing on the ceiling for a guide in making the cutout. Enlarge the pilot holes to 3/8 inches, working from below with a hand drill or very slow electric drill. Then make two more holes of the same size at the points representing the locations of the other two corners, as well. You now have entry points for a slender, pointed keyhole saw to get you started in making the cutout. Cutting the existing ceiling material out of the way is done from below, as there is less chance of shattering plaster if the cutting teeth bite toward the finished surface instead of from the other side.

Before beginning the cut, use a chisel and hammer to just barely cut through the finish coat if the ceiling is plastered. This is suggested because with plaster, minor penetration of the finished surface prior to cutting aids in preventing the shattering of surrounding finished areas to such an extent that the trim plate of the heating unit won't cover the ragged edges. If the ceiling is of drywall material, this prior nicking step will not be necessary.

Use of the hand-held, hand-operated keyhole saw is superior to that of the electric sabre saw in making ceiling cutouts for bath lamps, as the slower speed at which the work is accomplished tends to do less damage to abutting areas of the ceiling. Moreover, control is better, for you can tell when you encounter a pebble in the rough coat or rock lath, and you can then angle the saw backwards and forwards to help clear it. The electric sabre saw would have a tendency to grab it and snatch away attendant ceiling material.

Cut the short sides of the rectangular opening first. Then, cut down the first long side to within two inches of the shorter cut. Working from the opposite direction, cut down the other long side to within two inches of the opposing shorter cut. This leaves the relatively heavy piece of plastered ceiling material supported by two 2-inch strips of ceiling material on diagonally opposed sides, preventing premature fallout and rip-away of nearby finished surfaces. Supporting the piece to be removed with one hand, now carefully cut away first one of the two remaining 2-inch strips and then the other one.

There may be tabs protruding from the long ends of your housing, approximately halfway up its height, to permit nailing to the side of the joist. This is why we scribed the outline of the housing on the ceiling using the top of the housing as a guide. The lower edge probably has a lip on it, designed to pull up against the finished ceiling when the housing is inserted from below. But the tabs may prevent insertion from below. Make two overcuts longer than the dimension of the long side on which the tabs appear to permit them to be pushed through, allowing the unit to be recessed into the ceiling from below. Have someone hold it up against the ceiling while you go to the attic to nail it into place.

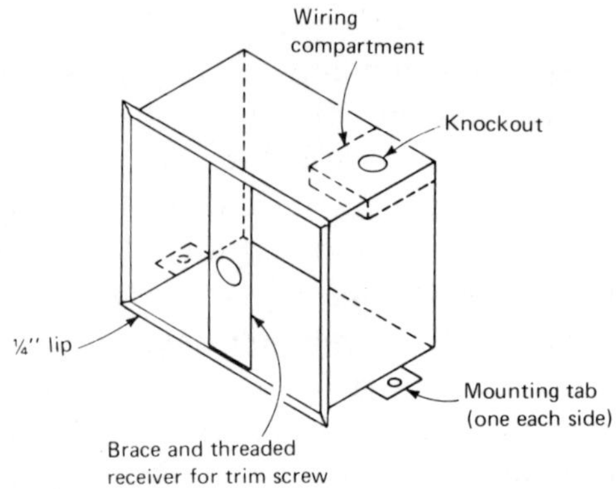

Figure 52: *Typical Dual Heat Lamp Housing*

It may pay to insert a cable connector into one of the knockouts in the top before securing the unit in place, as this is often more easily accomplished when all sides are accessible before mounting.

Figure 52 shows the probable location of a knockout in the side of the housing. This affords entry for our cable into the wiring compartment accessible from below. A removable cover, secured from below, permits finished-side terminations and assures accessibility to our connections in accordance with requirements that no joints in wiring may be inaccessable. Porcelain or ceramic bulb holders are not shown, but would be centered in the top of the unit in their respective sections.

Two 250-watt bulbs, comprising a load of 500 watts will be utilized in the heat lamp unit, in addition to a small internal fan for cooling, which may add another 15 watts. Although this constitutes a load of slightly less than 5 amperes when operating 115 volts, no special wiring on a separate circuit should be required. This circuit may be tied into the existing bathroom lighting circuit if the latter can be picked up in the attic. Set a junction box wherever you can intercept the cable that is carrying the circuit, and extend a cable from it to the heating unit, passing through the joists, not over them. Then, a leg must also be routed to a switch box located in the bathroom, perhaps adjacent to the existing light switch. If there is a neutral in the existing switch box, we may prefer to pick up our circuit there for the heating unit.

It is not desirable to use a standard light switch for control of the heating unit. We need to be able to assure automatic shutdown, should the heating unit be energized and then left on for any reason. For this reason, a timer is suggested, and in some localities it may be mandatory. Zero to 15 minutes is frequently the time span of the timer switch selected. It is generally installed in a deep switch box, and when turned on, it automatically ticks out the interval and turns itself off when the time is up. Take care to select a timer switch that is normally open, so that its internal contact is closed whenever the switch is turned, but spring-returns open at the end of the timing cycle. Your electrical supplier undoubtedly knows just the right timer to provide you with for control of the heating unit.

In the junction box in the attic, assuming an attic feed, circuit hot black will be connected to the white insulated conductor down to the timer, and black return will be connected to the black insulated wire serving the heating unit. Then white neutral will be connected to the system circuit neutral. This method gives us two-position control of the heating unit, either off or full on. Were it desired to operate the heating unit in two stages, 3-conductor would have to be extended to a combination timer and single-pole double-throw hi-lo selector switch. We would feed down white, and come back red and black, mating to red and black going to the heating unit along with a neutral. Preferably, the hi-lo selector switch would have its off position located in the center, so that when the toggle was there, the heating unit would be de-energized and the timer would be locked out. In Figure 53, we make use of the three-wire cable extended down from the fixture into the wall on which the manual 0-15 minute timer and hi-off-lo selector switch are located. Notice that the timer is wired first and that it isolates the selector switch whenever it is open. Turning the timer dial to energize the heating unit passes power to the double-pole switch, which is operated by a single toggle. The switch sections move in the same direction simultaneously whenever the toggle is operated. In the center position, no circuit is established. When the contacts make to the left, both lamps 1 and 2 are lit. But when the contacts make to the right, only one lamp, lamp 1, is lit, so that the heat-lamp unit

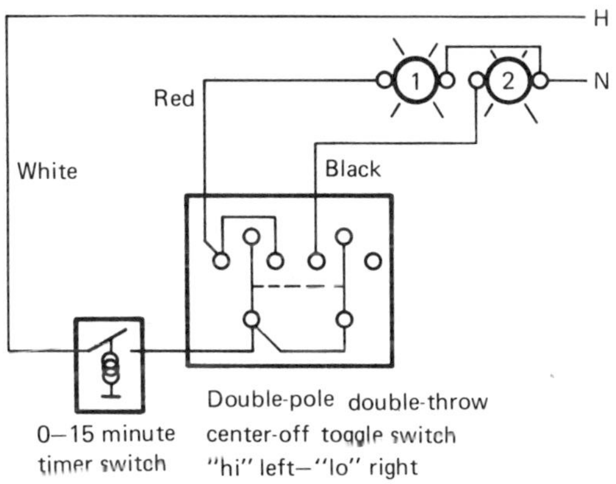

Figure 53: *Two-level Control of Dual Bulb Heat Lamp Fixture*

will operate at half capacity. Note that we connect white hot and feed down to the timer and switch location, and red and black become outgoing switched legs back into the fixture.

LIVING SPACE VENTILATION

An alternative to air conditioning can often be found in ceiling mounted exhaust fans, which induce air flow in through windows and deliver the exhausted air being replaced into attic spaces for relief through gable louvers or roof vents. These fans usually feature backdraft dampers, visible from the living areas, and mount horizontally in the ceiling, with the fan and motor in the attic. Their use might be especially effective when outdoor air temperatures begin to moderate enough to warrant inducing the outdoor air inside for cooling, such as in the evening. At such a time, mechanical cooling could be dispensed with, and outdoor air could be substituted for ventilation and cooling.

These fans differ from attic ventilation fans, discussed earlier, in that their primary purpose is living space ventilation, whereas attic fans are generally employed for reduction of attic temperatures in an attempt to reduce the radiant ceiling load that exists when an attic is quite hot, and the high temperatures migrate through the ceiling structure and radiate to the area below. As a result, ceiling fans for space ventilation are much larger, since they must move greater quantities of air.

A typical ceiling ventilation fan could have an electrical motor of 1/3 hp with a current draw of 7.2 amperes on a 115-volt circuit, practically half the rating of a 15-ampere circuit. Compared with a 1/20-hp attic fan drawing perhaps 1.2 amperes, the larger space ventilation fan constitutes a significant load, which draws our attention to two Code considerations.

First, it is usually convenient to connect the fan into an attic lighting circuit, which constitutes a general-purpose branch circuit. And Article 210 of the Code allows the size of permanently wired appliances on general purpose branch circuits to be no greater than 50% of the circuit rating. Therefore, a special circuit would not have to be installed to serve our space ventilation fan. If it did, it could have necessitated extension of a new circuit from the panelboard up into the attic area.

Second, Article 430 requires separate overload protection for the motor if the fan is to be automatically started and is over 6 amperes. This would call for a manual motor sentinel with integral heater, to sense motor current and open the circuit if motor current exceeded 115% of name-plate rating. Such a motor sentinel, which would require manual reset, would best be located where it is easily accessible, as it could also be the master on-off control. Additionally, the heat attendant to an attic location might cause premature tripping of the motor protector, creating nuisance shutdowns.

The motor protector, or sentinel switch, is important, because it has the ability to recognize when the fan is stalled. Such a condition would occur if the blades of the fan were jammed for some reason, and if it were sustained, stalled-rotor current draw would result in damage to the motor. Further, abnormal loading of the motor, as might occur if the fan blades dragged, could be detected as well, triggering a reaction and thus saving the fan motor from damage.

A hall location for the manual motor sentinel is ideal, as it affords on-off control for the ceiling fan and insures that ambient conditions surrounding the motor sentinel will not cause premature shutdown when the fan is running. Motor sentinels can mount flush in the wall just the same as a general-purpose light switch. A single-gang switch box will accommodate the sentinel nicely and can usually be recessed into a partition through the use of an "old work" box that relies upon void space to allow brackets to hold it in place.

The motor sentinel can include a neon indicator to show when power to the fan is switched on. It can be used in conjunction with a two-speed selector switch and thermostat as well, to afford the versatility required for the most flexible operation of the ceiling fan to meet individual cooling needs. Routing of the circuit should go to the motor sentinel first, then to the thermostat, and from there into the two-speed selector switch. Then the two switched legs from the two-speed selector would be extended on up to the two-speed fan motor for final connections.

ADDING KITCHEN RECEPTACLES ON A SEPARATE CIRCUIT

Remodeling invariably involves an increase in the size of the electrical service equipment, because more electrical equipment is usually added to the premises as a part of the remodeling project. The house is said to be rewired after a service change is performed, and this relates to replacement of the service conductors and the addition of branch circuits allocated to serve strategic areas and pieces of equipment. However, removal of existing wiring within the building rarely occurs as the original wiring, when properly protected against overcurrent, is usually suitable to remain in service for general lighting and general purpose receptacles. If more receptacles are to be added for general use, an existing circuit might be modified so that it serves fewer rooms with more outlets. The remainder of that original circuit would be fed separately from a new circuit, and additional general purpose outlets would be added in its area as well.

It is not feasible or necessary to replace all the wiring within the concealed construction of the residence; inspection usually reveals its condition to be satisfactory, so that better circuiting is the answer instead. And these new circuits should be routed to portions of the home where electrical load would be concentrated, thereby removing from existing circuits loads which could have been causing outages in parts of the home as fuses or breakers reacted to high current that occurred when certain loads came on the line.

Kitchens are a major area where recircuiting is particularly effective. Older homes might have just one or two receptacles in the kichen, but they could well be on the same 15-ampere circuit as the living room and dining room. So if wiring is being done during remodeling, most localities will require that a separate 20-ampere circuit be installed to serve the kitchen receptacles. Then, items such as the refrigerator, electric frying pan, and waffle iron can be placed in operation without penalizing other parts of the electrical system in the home. Dishwashers should be wired independently on their own 20-ampere circuit, because quite often they incorporate a 1,500-watt heating element that comes into play during the drying cycle, and this is a load of 13 amperes in itself. And, of course, the clothes washer must have its own 20-ampere circuit also.

Let us now examine the work often associated with the installation of new receptacles in a kitchen that is being remodeled. Wall construction and access combine to dictate methods to be employed in installing new circuits to new kitchen receptacles. Unlike other parts of the home where the requirement for an additional receptacle can be met through the use of surfacemounted raceway from an existing receptacle to the new one, our kitchen receptacles must be wired back to our service panel in order for separate circuiting to be accomplished.

Where new counters for the kitchen are to be set against outside walls, solid masonry construction may prevent fishing in wiring from the basement or crawl space below. Even if the wall was furred for lath, the shallow open space together with the blockage created by plaster protruding through the inside of the lath will undoubtedly preclude fishing in the wiring. But where counters are to be set against the wall, walls can be opened below the counter line, as cuts in the wall will be concealed once the counters are there. When the homeowner understands that the openings made for electrical work won't show, he normally has no objection to them. Then, outlet boxes can be mounted against the masonry construction, and if necessary, the block or brick can be cut back some to provide additional depth in order that the box edge be flush with the finished wall surface.

Hollow-core masonry units permit recessing outlet boxes into the voids between webs in the block construction. A hole can be knocked in the kitchen-side of the block with a hammer to allow the back of a switch or receptacle box to be recessed so that the front edge can be flush with finished wall construction.

Solid brick walls present more of a challenge. If they are furred, shallow boxes can perhaps be used, but keep in mind the minimum cubic inch space requirements for wires within the box discussed in Chapter 1. Normal cable-routing practices result in touch-and-go connections of four insulated conductors within the box. Then allowances for the grounding conductor, the device (receptacle or switch), and cable clamps inside the box, adds up to seven, and seven #12 wires require 15.75 cubic inches of space. A 3 x 2 x 1½-inch deep box will have less than 9 cubic inches, so an alternative needs to be developed. Let's change our box selection to a 4 x 2 x 1½-inch deep box that offers slightly less than 12 cubic inches, and let's also, omit cable clamps. Instead, we will install a cable connector in one of its ½-inch knockouts. Additionally, we will not route two cables to the box, but rather will use junction boxes in the crawl space or basement ceiling space and merely rise up with one #12 cable at each receptacle outlet location. Our cubic inch requirements, now based on two insulated conductors, one grounding conductor, and one device total just 9 cubic inches, so now our shallow box is large enough. Perhaps the box can be mounted directly to the brickwork with expansion shields and screws. This alternate method allows us to get by with the shallow box and avoid having to chip out solid brick to accept the depth of a larger box.

In Figure 54, the receptacle is depicted just above the backsplash on the kitchen counter. Cable clamping is accomplished outside of the box with a connector in lieu of internal clamping with built-in cable clamps and is visible due to the opening we made in the finished wall surface.

The junction box in the crawl space below requires grounding due to its location in a potentially conductive area. The damp earth in the crawl space makes a person working underneath the house an attractive path to ground, and a fault occurring within the junction box might energize the surface of

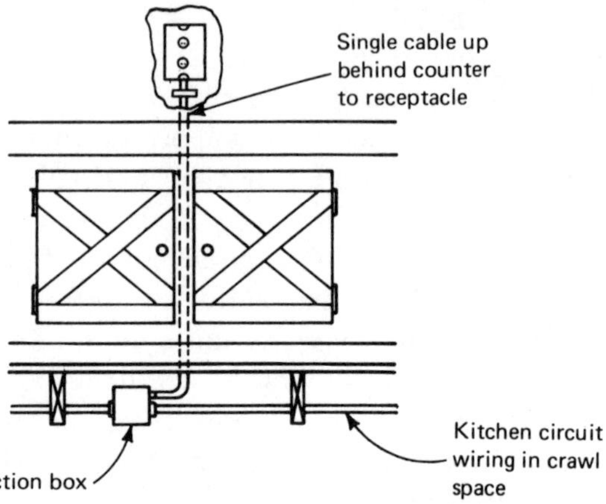

Figure 54: *Kitchen Circuit in Crawl Space*

the metal enclosure, offering a great hazard to anyone who might come in contact with it. For this reason, the grounding conductor needs to be bonded to the metal box by an approved means of bonding. This could be a grounding clip, approved for the purpose of securing the grounding conductor to the box or the bonding screw provided by the box manufacturer. The operation of the branch circuit overcurrent protective device would then be facilitated in the event that the energized conductor within came in contact with the box proper.

Special-purpose receptacles that must be added in kitchen areas for washer and dryer can probably be installed surface-mounted, with their wiring enclosed in EMT, extended straight down through the floor. Their location behind the pieces of equipment they serve assures us that the surface-mounting techniques employed will not detract from the appearance of the remodeled kitchen, when it is desirable to have our clothes washer and dryer in the kitchen area. It also saves us the necessity of having to install recessed outlet boxes flush in the wall. Figure 55 shows an elevation view from the side of such a surface-mounted receptacle.

Concealment of the EMT conduit and surface-mounted receptacle is afforded by the appliance in front. Clearance behind the appliance for installing exposed wiring is frequently available due to plumbing and venting connections, which necessitate standoff. Here again, be sure to bond the metal box and EMT to the circuit grounding conductor to insure protection against electric shock in this potentially conductive location around the washing and drying equipment. Of course, the washer requires a 20-ampere 115-volt circuit, and the dryer generally needs a 30-ampere 230-volt circuit, with matching receptacle.

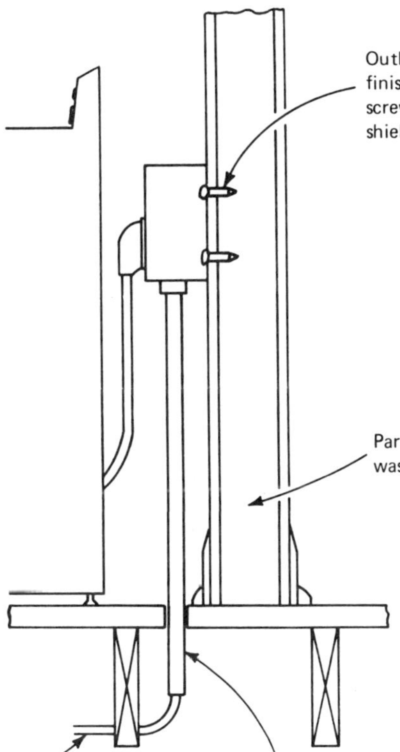

Figure 55: *Surface-mounted Washer/Dryer Receptacle in Finished Area*

INSTALLING ADDITIONAL OUTSIDE RECEPTACLES

The convenience of having receptacles outside the house where you need them helps reduce the use of extension cords outside the home. Hedge clipping is much easier when receptacles are strategically located on the outside walls of the house to facilitate the minimal use of long cords. Being able to plug in a service light on the outside when it's necessary to work on the car makes the job much easier. And for those with electric lawn mowers, exterior receptacles mean greater flexibility in being able to reach all areas of the yard.

Newer homes feature at least one outside receptacle because of the Code requirement for it in Article 210. But older homes generally won't have any unless they have been added by previous owners. Frequently, it is desirable to install one or more outside existing homes and tie them into existing lighting circuits. But remember that Article 210 requires ground-fault circuit protection for receptacles that are installed out-of-doors. So the receptacle itself must be of the ground-fault type, or a ground-fault circuit breaker must be provided at the service equipment. The latter might be preferable when two or more outside receptacles are to be installed and tied into the same circuit, or have a special circuit run to them. The special characteristic of ground-fault circuit protection is that opening of the circuit occurs before anyone can receive a lethal shock in the event of a fault-current to ground.

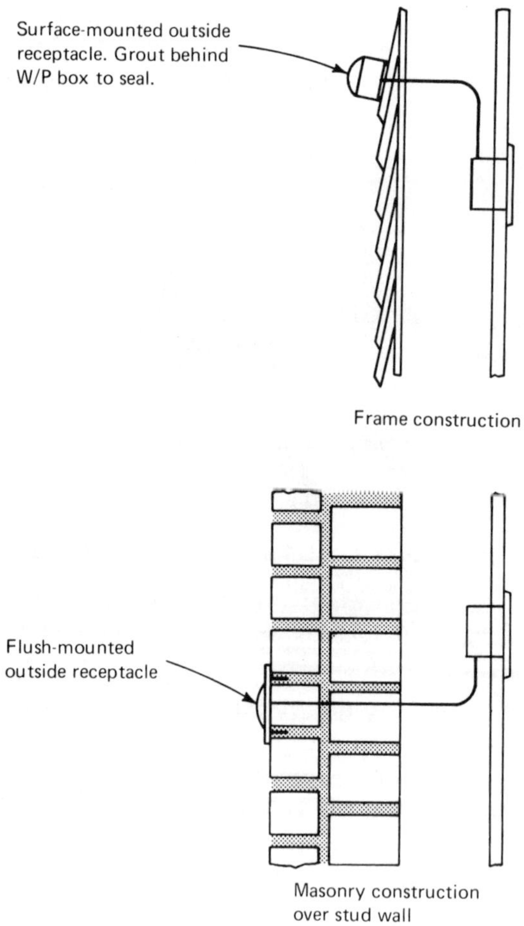

Figure 56: *Mounting of Outside Receptacles*

The ground-fault interrupter receptacle or breaker is set apart from conventional types by its ability to distinguish between normal current flow and fault current flow. An integral comparison network monitors current flow in both the hot and neutral wire of the circuit, and any imbalance causes the opening of the circuit passing through the protector. An accidental fault flow of current from the windings of a power tool to the surface of the tool and thence through people or the branch circuit grounding conductor to ground is an example of the type of ground-fault that would trigger reaction of the protection network.

Two alternatives in mounting the outside receptacle present themselves: flush-mounting or surface-mounting. In surface-mounting, a weatherproof receptacle box is used in conjunction with a weatherproof receptacle cover. But in flush-mounting, a standard receptacle box is recessed into the wall with its outer edge flush with exterior finish construction, and a weatherproof receptacle cover forms a seal against the house exterior. Either alternative is suitable. Surface-mounting is easier in existing construction, but we will examine both methods. Flush-mounting is often preferred because it offers minimum projection from the side of the house, even though its installation requires more work and greater care in cutting the outside wall.

In the top illustration of Figure 56, we see a surface-mounted weatherproof box against the clapboard of a conventional frame home. A simple ½-inch hole permits penetration by the electrical cable into the back of our outside receptacle box. The box with gasketed cover in place is weatherproof, and grouting behind the box before anchoring it fast with wood screws insures no seepage of water down behind the back of the box.

Flush-mounting is shown in a masonry wall featuring brick veneer over 4-inch masonry block with stud furring on the inside. Half of a brick is cut away to permit recession of the rough-in box. Then, a ½-inch hole is bored through into the inner wall. Both kinds of installations shown take advantage of back-to-back installation of the outside receptacle with an existing inside receptacle. Cable runs are minimized, since only enough is required to extend the existing circuit through the wall into the new outside receptacle. By taking careful measurements, you can judge the suitability of various outside locations that line up back-to-back with existing inside receptacles.

Removal of the half brick should be accomplished with minimum percussion disturbance to surrounding masonry. A ¼-inch masonry bit in a hammer-drill can be used to drill holes to cut the brick in half. Then, a 1-inch masonry bit is used to cut a round hole in the center of the piece to be removed. This allows the piece to be removed to cave in upon itself, as you begin to apply hammer and brick chisel. A 2-pound mall and a 14-inch brick chisel with a 1-inch-wide cutting edge work nicely. Cut the webbing left

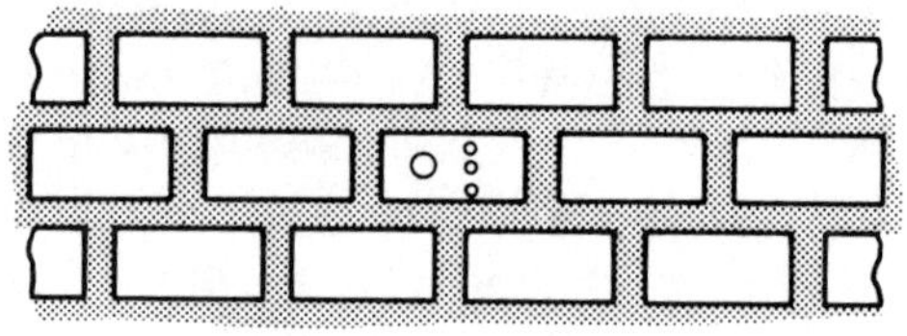

Brick is drilled with masonry bits to divide the brick into two pieces, and allow it to cave in upon itself.

Figure 57: *Brick is drilled with masonry bits to divide it into two pieces and allow it to cave in upon itself.*

from drilling out the ¼-inch holes. Then, chisel the brick so that crumbling occurs in toward the center. The 1-inch hole permits the brick to break inward, causing minimum disturbance to mortar and surrounding brickwork. When the half if removed, you'll see the masonry block, backing up the brick veneer. Now drop down to an 8-inch brick chisel with ½-inch-wide cutting edge to square up the hole with clean edges to allow the box to slide back just into the hole. Drill the ½-inch penetration hole for the cable next.

The box may be anchored with #12 or #10 x 1-inch-long pan-head self-tapping screws and ¼-inch x 1-inch-long plastic anchors, installed horizontally. A ¼-inch hole must be cut through the ends of the metal rough-in box, and its relationship to the brickwork on either side should be marked with a pencil. Then, a ¼-inch masonry bit can be used to drill holes for the anchors. Once the box is then slid permanently into place, it can be firmly anchored to prevent movement.

Sometimes, it is difficult not to disturb surrounding brickwork. Any openings around the sides of the box can be filled with patching concrete or ductseal before the weatherproof cover is installed. If the mortar at the top of the box breaks away, use the ductseal (available at electrical supply houses) to form a waterproof collar across the top, which will compress slightly when the weatherproof cover is put in place. This will insure that no water enters the box. Remember, the slightest fault-current that occurs as a result of water entering the box will trip the ground-fault protection device, and the receptacle will be out.

Installation back-to-back with existing inside receptacles can usually be best accomplished if the outside receptacle is positioned slightly above or below the inside one. This is because fishing the connecting cable in through knockouts is easier with the receptacles arranged that way than it would be with the boxes facing one another, back-to-back, through the wall. Of course, the inside receptacle must be removed and its circuit de-energized. Once the holes for the cable are cut through, use a light on one side to illuminate the passageway, visible from the other side. This will help immensely in assisting you to guide the cable where you need it to go. Don't forget, a pull wire can also be used to fish it through.

FLOODLIGHTS

Illuminating the yard with light is important at night when emptying the trash or making a trip to the tool shed. It's very easy not to see an object that is irresponsibly left in your path, and then stumble over it and trip and fall. Proper lighting of the outlying areas around your home is easily accomplished through the use of floodlamps strategically located on the upper overhanging portions of the house. Most any convenient branch circuit can be utilized to serve the lights, as two 150-watt bulbs add only 2.6 amperes to the circuit on which they're connected.

Conventional floodbulb holders permit angular adjustment in universal joint fashion. In the popular two-bulb fixture, each arm offers 360° rotation at its base as it threads into the baseplate. Locknuts can then be tightened to secure the arm at the point desired. This is followed with elbow adjustments to each arm, held firm at the right spot with tightening of an elbow screw that engages mating teeth on each side of the adjustable joint. Wiring is concealed within the arm, and a gasket makes the seal between the base and the weatherproof mounting box raintight. Each floodbulb then screws into the porcelain socket within the arm and is flashed with a field-installed or fixed-bulb gasket to forbid the entrance of water around the bulb. When purchasing floodbulb holders, note that the fixture features porcelain bulb holders, as many of the bulbs rated at 150 watts warn against use in other than porcelain bulb holders due to high temperatures generated by the floodbulb.

Use of flood lights outdoors must be reserved to fixtures approved for the purpose. Do not attempt to use a standard interior lamp socket outside, as its construction does not render it safe for outdoor use, and safety hazards can result.

Popular locations for the floodlight are under the eaves of homes, where the fixture can be mounted to the underside of the overhang. This spot is usually accessible from the attic, and cable can be run into the void in the eaves from within the attic space. Gable mounting finds frequent approval also, as it gives the fixture height for illumination of a broad area. Through-the-wall penetrations in the attic allow extension of branch circuits to serve gable locations. Perimeter wall switch locations may require installation of conduit on the outside of the home if access to the desired switch location cannot be found in the interior.

If your option is for more than two floodlamps, it is important to keep in mind that each bulb constitutes a load of 1.3 amperes. Dual lamps on each of four corners of the home total 10.4 amperes, and difficulty is likely to arise when this size load is randomly connected into an existing branch circuit. In normal use, when the circuit is only partially loaded, energizing the floodlamps can raise branch circuit current flow to an amount greater than the overcurrent protective device in the panelboard might handle. Special consideration needs to be given to placing that many lamps on a single circuit by themselves.

Floodlamps designed for outdoor use are intended to endure the temperature extremes and weather conditions likely to be encountered throughout the year. Note the extra thickness of the glass from which the floodbulb is constructed. Conventional interior bulbs that are exposed to cold rain when they are hot from use can be expected to break, but the floodlamp will remain intact through all seasons.

Mount the weatherproof box against the house using screws sufficiently long to insure solid mounting. In masonry, blue plastic anchors and #12 x1-inch pan-head self-tapping screws work nicely when installed in ¼-inch holes. In frame, this same screw will take hold well in a 3/32-inch hole. In manufactured siding, it is advisable to use a #12 x 2-inch-long screw through the siding and into a stud or partition top plate to guarantee solid anchoring.

Undoubtedly, you will use an extension ladder to reach the desired installation spot. Make sure the ladder is in good condition. The ladder should be out from the house at the base one-quarter the distance extended. If the ladder is extended a total of 16 feet, it should be 4 feet out from the perpendicular at the ground. If it's up to 20 feet, it should be out 5 feet at the ground. These recommended settings insure minimum deflection of ladder rails when climbing, while affording proper angle of lean. Do not lean over to reach while up the ladder. If you can't reach your work without leaning, come down and move the ladder. Do not exceed the ladder manufacturer's warnings regarding overextension of the ladder, as this weakens the ladder's resistance to deflection. Take only as much up the ladder as you can easily manage. Segment your work, and make multiple trips up the ladder as required.

RELOCATING FIXTURE OUTLETS

Sometimes it is desirable to relocate a lighting fixture from one location in a room to another. Consider a home that is sold to new owners who decide that their preference is for the bathroom lighting fixture to be in the center of the bath ceiling instead of over the sink. The room is a half-bath, and entry to it is from a hall, with the light switch just to the right as you come through the door. The home is several years old, so walls are in place and wiring is concealed. Finished space is above and below the bath, so access from unfinished areas is out of the question.

From a practical standpoint, we are interested in running as little new wiring as possible, with minimal disturbance to existing finished surfaces. Some patching and painting is going on anyway, since the new owners are redecorating. But even if they weren't, we are interested in keeping patching to a minimum. Since cables pass through wooden framing members when they run at right angles to them and are stapled to framing members when they run with them, rerouting of existing concealed cables without extensive wall damage is frequently impossible.

In examining the floor plan layout of the bath in question (Figure 58), we note that the new location of the lighting fixture outlet will be closer to

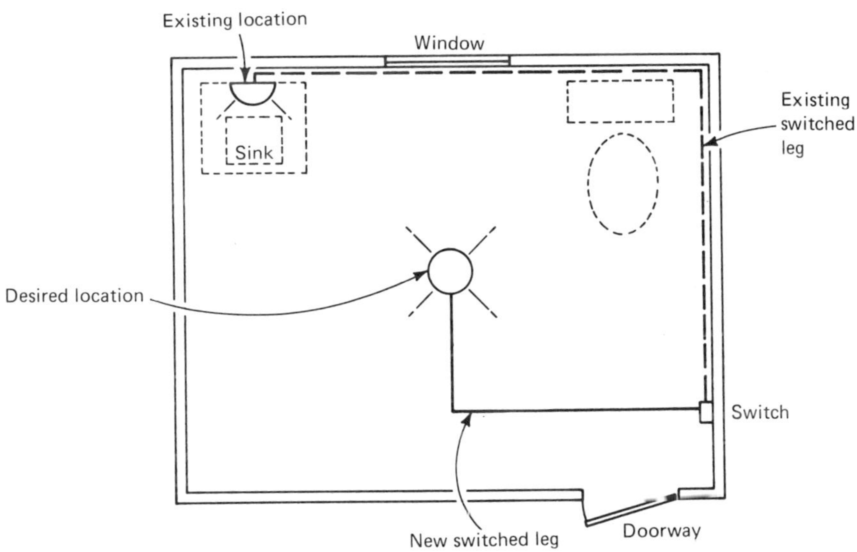

Figure 58: *Relocating Fixture Outlet*

the switch than the outlet over the sink was. This, plus the fact that the old switched leg cable running over the window is irretrievable, suggests that abandonment of the old cable and installation of new cable might be more feasible than attempting to reroute the existing cable. Plans are to plaster over the opening left in the wall where the existing fixture is currently located, provided that only the one cable comes into the outlet.

Disconnect the line to the existing light at the switch box, and push it out of the box so it cannot be reconnected. Remove the outlet box over the sink, and push the abandoned cable back into the wall if the feed is to the switch box and no other cables enter the fixture outlet box. Now the abandoned cable is both dead and irretrievable.

Locate the underside of a ceiling joist closest to the center of the bath ceiling. Holding an octagonal outlet box to the ceiling at this point, scribe its outline on the ceiling, and cut out a plug to permit recessing the box into the ceiling material. Check whether any additional depth is necessary to allow the edge of the box to be flush with the finished line of the ceiling. Use a 3/4-inch wood speed bit to cut away enough of the exposed lower edge of the ceiling joist to accommodate the octagonal outlet box, and secure the box in this notch with #10 x 3/4-inch pan-head self-tapping screws after installing a cable connector in one of the side knockouts of the box.

Use a 1/16-inch hex key tool with a handle as a pilot punch to locate the remaining joists along a line representing the path to a point in the ceiling just over the switch box, and make thumb-sized incisions across the lower edge of each joist so that our new cable can be fished from point to point and stapled to the lower edge in a notch cut just deep enough to accept the cable. At the corner, where the ceiling meets the wall, cut two incisions at 90° angles to each other, exposing the edge of the partition's top plate and the underside of the abutting nailer. Notch both of them with a sharp chisel, and staple the cable into the corner. Then fish the remaining cable length on into the switch box, and our new switched leg is roughed in. Install nailstops to protect the cables. Finally, the new fixture can be hung and the line can be reconnected in the same way that our old, abandoned line was. Patch all openings with patching plaster.

If the fixture outlet over the sink had contained the feed for the light, three cables would probably have been present, hot in, hot out, and switched leg, which would have run over to the switch. In such a case, retention of the switched leg would be a better choice, because the old outlet box, representing a splice point for the circuit passing through, could not be removed and done away with. Instead, a decorative blank-off cover could be utilized after a new line was run the short distance to the new fixture outlet. Of course, this could have been done in our first example, too, but the desire there was to remove the old outlet box, if at all possible, and patch the opening left by it.

In either case, the work, if carefully performed, results in a minimum amount of patching and painting and accomplishes our objective of moving the outlet for the bath fixture to a more desirable location.

INSTALLING LIGHTNING ARRESTERS

An excellent buy for the money in today's market is the lightning arrester, designed to protect the home's internal wiring system against surges in electrical voltage as a result of electrical disturbances associated with summer storms. During a lightning strike, excessively high voltages can be induced into the utility company's distribution and service entrance conductors, and they may find their way into the home, unless preventive steps are taken. The lightning arrester will drain excess voltage to ground and keep it off interior wiring systems.

Such voltage surges have been known to destroy filaments in light bulbs that happen to be burning at the time they occur and damage electronic components in televisions and stereo systems. Range ovens and electric heating may come on after a close lightning strike when the contact points in thermostats controlling them are welded together by the excessive voltage as it attempts to jump the gap.

The lightning arrester, installed at the service entrance point, offers a sufficiently high resistance to normal operating voltage, so that no carry-off occurs. But when the voltage rises significantly, as it does when lightning strikes, the arrester permits the higher, undesirable voltages to pass to ground and be carried harmlessly away. Both legs of the incoming power, plus the grounded neutral, are connected into the arrester, so that it may monitor conditions on both incoming service conductors and dump to ground when necessary. Installation should be limited to qualified personnel only.

DOORBELLS

Signaling systems for arrival of people are low voltage and utilize smaller-gauge, lower-cost bell wire. Push buttons at front and rear doors allow selective announcement, so that callers can be immediately identified as being at either the front or rear door. Gongs, chimes, or tones, designed to create distinctively different sounds in response to push buttons operated at different doors, are frequently offered for selection by manufacturers. Low-voltage transformers for powering the bell circuit must be installed in an accessible location. Quite often they are placed in crawl spaces, attics, basements, or utility rooms. The most convenient 115-volt, 15-ampere circuit can be used, as power requirements rarely exceed 15 watts.

Voltages produced by these small transformers may be 10, 12, 16, or 24 volts. Most often, 12- or 16-volt models are encountered, and their voltage must match the voltage of the bell or chime being used. This is especially important in replacement work, where new transformers are to supplant old ones. In new work, the chimes are usually purchased with buttons and transformer, all at the same time.

In Figure 59, we see how a dual tone, 16-volt chime set should be wired. Just the low-voltage wiring is shown. The only line-voltage wiring would be at the point where the transformer is connected into our 115-volt household branch circuit.

The chime terminal strip will be labeled "front, transformer, rear," to designate the proper connection point for each push button and transformer. Each button completes a circuit to its respective plunger solenoid, striking the appropriate tone to announce someone at the front or rear door. The circuit is through the solenoid, to activate the plunger, and back to the other side of the low-voltage transformer circuit.

For ease of operation at night, each push button might feature back-lighting, making it easier to see in the dark. This is accomplished by connecting a low-voltage bulb within the push button across the terminals A-B. Thus, when the button is not depressed, a very small amount of current passes through the bulb, causing it to glow, and shine through the translu-

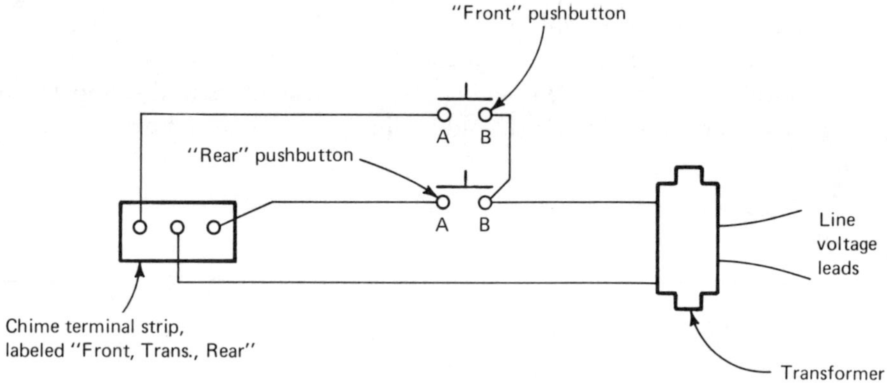

Figure 59: *Typical Connections for Dual Tone Door Chime Set*

cent button. The solenoid within the chime acts as a conductor, because of the extremely high resistance of the small bulb. However, insufficient current passes to energize the solenoid, so the chime remains silent. When the button is pushed, the light is extinguished, as current selects the path of least resistance through the contacts of the button. This permits the solenoid to draw sufficient current to energize, and the chime reacts audibly.

Push buttons are installed outside and so are subjected to the dampness and temperature extremes of the outdoors. This corrosive environment causes the contacts of the buttons to oxidize, which, when coupled with the low voltage of the circuit, may, over a period of time, cause the chimes to stop working. The buttons may be removed when this happens, and their contacts may be cleaned, to expose bright, shiny metal again, to restore operation. Very fine emory paper can be used quite effectively to renew the good current-carrying characteristics of the contacts within the buttons and once again permit the chimes or bells to operate.

Bell wire is frequently 18 gauge, 2 or 3 conductor, with a lightweight polyethylene jacket for protection. It may be installed like our non-metallic-sheathed cable through studs and joists. Usually a hole is drilled in the door trim, and the bell cable is pushed through to the outside, for later installation of the button. Proper precautions should be taken to protect the cable against mechanical damage, as the wiring is smaller and might be crushed or broken if it is exposed where there is lots of activity.

SECURITY SYSTEMS

Protection against intrusion has become a familiar part of many of today's home wiring systems. Such protection is accomplished with security systems that can monitor the home's potential entry and egress points and sound an alarm in the event that one of them is violated while the system is on guard. By energizing the security system and placing it in the secure mode, no doors or windows can be opened without audible alarm. The alarm would not only warn the homeowner against intrusion, but in all likelihood would act as a deterrent to the would-be intruder and cause him to leave rather than risk confrontation.

The basic security system that we will examine utilizes contact closure devices, activated by opening the door or window, to give an alert. Additionally, circumvention of the alarm circuit is precluded through circuit supervision that relies on all conductors being intact and makes outsmarting the

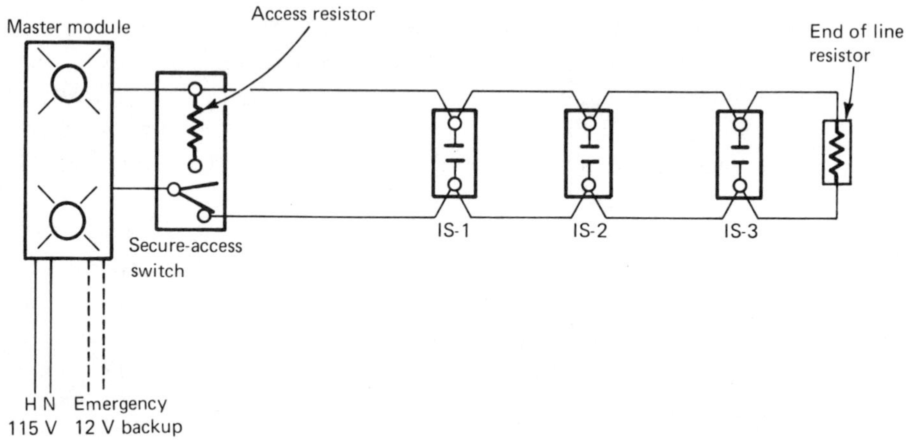

Figure 60: *Supervised Alarm Circuit with Open-Secured Contacts*

alarm circuit by cutting the wires impossible. Breaks in the wiring are also guarded against, by the same supervisory circuit that prevents deactivation through wire cutting. Either of these situations can be distinguished from a true alarm by creation of a trouble alarm, separate and distinct from the intrusion alarm, that would be discernible through differently colored visual reporting or differently coded audible reporting. Constant tone for mechanical or electrical trouble and intermittent tone for intrusion provide an easily recognizable distinction. Or, differently sounding tones could be utilized in the same fashion. Likewise, steady and flashing light could be used in lieu of two separate differently colored lights.

Figure 60 shows the kind of security system envisioned. In it, the whole house could be treated as a single zone, with all alarm contact devices on the same loop; or, various portions of the home might be zoned separately, with a loop brought back to the master module for each zone. This latter would aid in localizing an alarm in the event of intrusion. Note that normal 115-volt power is brought to the master module, which incorporates an internal transformer and rectifier to convert the incoming housepower to 12-volt D.C. for use in the alarm loops. Additionally, a good system will include back-up power, so that in the event that normal power is lost, the alarm system will remain active. An emergency power pack is generally available to come into play whenever regular sources are lost. It includes a charging module to keep the battery source properly charged as long as commercial power is available.

The alarm loop is extended to contact devices that open when doors and windows are closed. These devices might be jamb switches, or merely magnetic proximity switches. No circuit is made across any contact device until a window or door is opened. The security sensors are said to be normally closed; they are spring-loaded to make contact in their free state, i.e. when not held open by a door jamb, window sash, or magnetic attraction from a mated proximity magnet. In the schematic of Figure 60, we depict the sensor contact devices in the secure mode, or held open by the presence of the closed doors and windows. In this mode, and with the secure/access switch in the secure position, a continuous circuit is formed through all the wiring by means of an end-of-line (abbreviated EOL) resistor, and this resistor permits only enough trickle current to flow in the wiring to satisfy and silence our trouble-signal emitter in the master module. In this status, the circuit wiring is said to be supervised, i.e. monitored as to the integrity of its electrical loop. If an "open" develops anywhere along the loop, our trouble signal will sound, advising us that the integrity of the loop has been violated. This is important, because were we not to be made aware, a break between intrusion sensor IS-1 and IS-2 would prevent an alarm at IS-2 or IS-3 from being received.

Opening a window or door when our system is in the secure mode would close a contact device, which in turn would create a bypass circuit around EOL, permitting the alarm relay in the master module of the zone to pickup, and sound the alarm. Should legal access be desired, the key-operated secure/access switch would be placed in the access mode. In so doing, contact is first made with the access resistor in the key switch to inhibit sounding of the trouble alarm. This is brought about by continuous movement of the key toward access, which opens the loop out through our field contact devices to prevent contact closure from being detected by the master module.

Because opening the loop violates the supervisory condition, our key-switch resistor is matched to the EOL so that it appears identical to the master module. This assures that false trouble signals are inhibited when key-switch cams isolate the alarm contact devices in a one-two sequence as the key is turned to access. Turning the key back to secure restores circuit integrity before the internal matching resistor is removed from the line, forcing the master module once again to check the entire loop for supervision of line integrity. Indicators are frequently provided for, both inside and outside the home, to tell us at a glance whether the state of the intrusion alarm system is accessed or secured. A test button is oftentimes built into the master module to permit periodic validation of annunciation capabilities. A system of this type is available through most electrical supply outlets as a product of one or more manufacturers whose hardware may vary slightly, but basically the principle set forth here can be said to be prominent in the home security field.

If it were your desire to economize and still enjoy the comfort of home security, you might opt to purchase the contact devices only of the normally open variety. These security sensors would be held closed in the secure mode and would open the circuit if a door or window was opened when the system was on guard. You could then utilize a mere self-powered relay to effect home security. 'Self-powered' pertains to the incorporation of a transformer within the relay itself for voltage reduction. Let us now look at such an economy-model system.

Figure 61 shows a single-pole double-throw relay with 24-volt coil and built-in 115/24-volt transformer. The contact configuration is common-to-normally-closed when the relay is de-energized and common-to-normally-open when the relay is energized, as it is shown in the figure during the secure mode. All alarm contact switches are held closed by the door jambs and window sashes being shut. The two-position, manually-operated toggle switch S-1 is labeled "secure" in the open position and "access/reset" in the closed position. One wire is wired in series through all the field contact alarm devices and terminates on the normally open terminal of the relay. When the home is first secured, S-1 is closed in the "access-reset" position. This holds the relay in, positioning the contacts as we see them shown, to keep our annunciator silent until doors and windows are closed.

Once everything is secure, S-1 is opened to its "secure" position. This forces the holding circuit for our relay out through field contact devices, which must remain closed in order to keep the relay energized. If one of

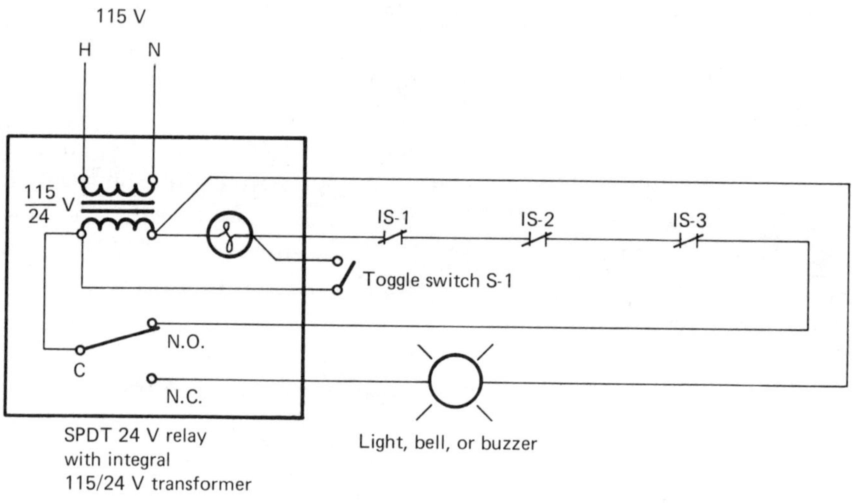

Figure 61: *24VAC Economy Intrusion Home Alarm System*

them is opened, the relay drops out, and contacts transfer current along the common-to-normally-closed path, placing power across the light, bell, or buzzer. S-1 must be moved to the "access/reset" closed position to extinguish the alarm and reset the relay. S-1 cannot be opened without the alarm sounding again until all field contact devices are again closed by shutting doors and windows. This particular version of home security does not distinguish between intrusion and trouble, sounding the alarm the same way in both cases. No reserve back-up power is provided either, meaning that in the event of a momentary power failure, the relay will move to alarm status, and restoration of power will sound an alarm until the switch S-1 is moved to "access/reset" to pick up the relay again. A ten-second time delay relay could be incorporated into the circuitry to forestall this condition if the momentary power loss was less than ten seconds. It would not be desirable to prevent the alarm from sounding after restoration for longer than this length of time, for that would allow what is perhaps too long an unprotected interval to exist in the system. But this inconvenience might very well be tolerable in light of cost savings over commercially available packages that would cost more than this simple version of home protection and intrusion annunciation obtainable with a minimum of components.

Wiring for the home security system is best installed when the home is under construction, but many systems are put in after the house is built. With walls and ceilings open, as they are in new work, installation of low-voltage wiring is so much easier. Later on, much fishing will have to be done in order to have a concealed installation. In some cases, low-voltage security wiring may have to be run exposed along door moundings or atop baseboard if getting into walls is impractical and opening walls is undesirable. However, the smaller wire used in alarm wiring is easier to work into wall cavities, and its extra flexibility may permit entry into some areas where entry is impossible with the larger cable used in regular house wiring. Wiring can be run overhead in basements and crawl spaces, with penetrations up into the walls at the appropriate points. For protection of the wire, holes should be cut through floor joists when runs are at right angles to them. This will help prevent their being snagged by service personnel who have to go under the house when the need for service arises in areas such as plumbing, termite treatment, and carpentry or masonry repairs. Be sure power wiring to the relay or master module is sized to compliment circuit overcurrent protection. Connection to a 15-ampere circuit is satisfactory because of the low power requirements of either type of security system discussed. Terminate the cable with an approved cable connector if cable installation is within wooden members, or drop to the relay in EMT or flexible conduit, using an EMT or flex conduit connector.

ADDING A SECOND FRONT PORCH LIGHT

So many of the lovely homes being constructed or redone today are patterned after homes that were prevalent during the colonial period of our country. You can see evidence of this throughout the United States, where cedar shake roofs with a 12/12 pitch, multiple fireplaces, and graceful verandas are combined in architectural designs that produce housing reminiscent of a period in our past that bespeaks of patriotism, charm, and warmth. Lighting on the outside of the home can also contribute to the desired early American effect when fixtures are selected that reproduce the soft lighting effects of lamps of old in situations where the intensity of modern-day lighting is not called for. It is frequently desirable to replace existing outside fixtures with special-effect ones that date back to colonial times and to feature two front porch fixtures, one on either side of the front door. Code requirements mandate only one fixture at entry and egress points of residences. This can be over the door, or on the wall beside the door. If a home has only one fixture in the ceiling of a small front porch, special installation techniques have to be applied in order to enjoy the lighting effects of dual wall-mounted outside fixtures on either side of the door. Consider a large two-story colonial home with a small front porch with one fixture in the ceiling. It is desired to feature a reproduction fixture on either side of the front door, where there are no fixture outlets now. The existing porch light switch is just inside the door on the wall on the strike side of the door, ganged with two other switches that control upstairs and downstairs hall lighting. The home is sided with manufactured siding and is well insulated. Our task is to install the two desired fixtures and conceal the wiring, as would be the case if the fixtures had been installed during construction of the home.

In examining this task, we notice that moulding is used to trim the porch ceiling around its edge. Careful measurements indicate that the moulding, installed to cover the joint where the porch ceiling and the siding on the house come together, is just below the top plate of the first-floor front wall partition. Our best alternative appears to be to conceal our cable behind this moulding strip, which is easily removed. This will give us horizontal run space, allowing penetration of the siding behind the moulding immediately above selected locations for each of the fixtures. We can then fish the walls from the under-moulding penetration to the point where our cable will emerge to serve each fixture. This will tie the two fixture outlet points together, and when the horizontal run is recovered with the moulding, no wiring will show. To forestall piercing the cable with nails when the moulding is reinstalled, we break away ½-inch of the edge of the drywall ceiling material to permit tucking the cable up out of the way. Joists in the porch ceiling run from side to side, and we find that the cable can be pushed up into the gap that we have created. A continuous pull of unbroken, unjointed cable is necessary, so we make the pull from fixture outlet to fixture outlet.

The electrician's fishtape must be inserted from the lower hole at the fixture location and pushed to the upper hole, later to be concealed by the moulding. The angle is too sharp at the opening near the ceiling to allow guiding the nose of the fishtape toward the other hole. But we desire to pull from the lower opening up to the upper one. So the fishtape is pushed in and retrieved at the upper opening with a coat hanger, which has been cut and converted to a straight wire with a hook rolled in one end. With this hook, we snare the fishtape and pull it out of the upper opening. Now, a piece of #14 wire is pushed into the upper hole and pulled out of the lower one, making a pull wire for starting the cable on its way to the other fixture outlet point. Enough to follow the path required, plus some extra slack is pulled through, but the cable is not cut yet. The fishtape is again inserted, this time in the other fixture outlet hole, and pushed to the upper opening on the other side. Our pull is in the correct direction this time, so the cable is pulled in and maneuvered out of the hole where the second fixture will go. We then work our way back toward the first hole, where the cable still emerges from the carton and enters the wall, and we take out slack and tuck the cable up into the crevice that we have prepared for it. Finally, our lineman's pliers are used to cut the cable, and there we have the two points tied together. After reinstallation of the moulding trim at the porch ceiling line, the upper holes are no longer visible, and all that can be seen is two pieces of cable protruding from two 1-inch holes on either side of the doorway.

The next item is extension of the ceiling fixture circuit to the outlet closest to the switches. After removing the multi-gang switch cover, we bring the switches forward enough to permit fishing a cable from the outside into the switch box for connection into the switched light circuit. A 3-gang plastic switch box has been used, so a flat 7/8-inch wood bit is used in a 3/8-inch variable speed drill motor to cut a round hole in the back of the box, the same size hole required for a ½-inch cable connector. The 3-gang switch box was nailed onto the door jacks, i.e., vertical studs that frame the doorway, so we have clear space behind the opening in the outside wall, except for insulation. Other cables to the switch box are stapled to the studding to meet the Code requirement for cable support set forth in Article 336. This way, loose cables in the void space do not hamper our attempts to install a short piece of cable between the switch box and the closest fixture outlet point.

The circuit serving the switch box is de-energized, and our fishtape is inserted through the new hole we have cut, angled toward the fixture. Then, the wire coat hanger is again used to snare the loop in the nose of the fish-tape, whereupon it is pulled through to the outside. Now, cable can be attached and pulled back into the switch box. After we have a minimum of 6 inches of slack outside and within the switch box, the cable is cut, giving us two cables protruding from the hole on the outside near the switch. Inside, a plastic press-fit cable connector is used to secure the cable to the switch box. Outside, two four-inch diameter round fixture boxes ½-inch deep with center back knockouts are fitted with ½-inch cable connectors to grip the cable and then secured to the wall surface with #12 × 1-inch pan-head, self-tapping screws. At the completion of this phase of the work, we are roughed in for the two new fixtures.

Each of the colonial fixtures incorporates a large, rectangular wiring compartment on the back of the fixture through which a 1/8-inch all-thread mounting nipple will be used to hold the fixture to the mounting strap that is to be secured to the shallow boxes. On the exposed side of the nipple, a 1/8-inch female threaded brass cap and decorative washer add the finishing touches. Cables are stripped, conductor ends bared, and electrical connections made before fixture mounting. Be sure to bond the grounding conductor to the edge of the shallow round box with an approved grounding clip.

Finally, we extract the gathered neutrals inside the switch box and snip off the twisted conductor ends to start with fresh ends and perhaps a larger Wire Nut®* to make a new joint that this time will include the neutral leg to the new fixtures. The grounding conductor within the switch box is connected to a grounding pigtail left by a previous electrician. Then, the black conduc-

switch, leaving the existing fixture leg connected also. Once the switches are all remounted and cover plate is put back, the ceiling fixture on the porch is removed and wires are capped off. Then, a removable cover plate is used to close up the outlet where the porch ceiling fixture used to be, still leaving it accessible. Once painted, it will barely be noticeable.

Thus, two wall-mounted fixtures are substituted for one ceiling hung fixture, and all wiring is concealed. Re-installation of the ceiling fixture can be accomplished at a later date, should that be desired.

A masonry home may have caused us to select another way to get the new wiring in to serve new fixtures. We could have made openings in the

*Wire Nut is the registered trademark of Ideal Industries, Inc.

inside wall, crossing studding with notches and nailstops as we made our way from one fixture outlet point to the other. A masonry hammer drill with a ½ inch long bit could be used to cut openings that permit through-the-wall penetrations for cables to the new fixtures.

Wiring to the new lights must match the existing circuit in current carrying capacity. Since it is likely that a 15-ampere circuit served the existing switch box, #14 copper should be satisfactory. Had #12 been used, our new cable would have to be the same size so that the integrity of the circuit's current carrying capacity is kept intact throughout the entire pathway of the branch circuit. In appearance there is no discernable distinction between the wiring thus installed and that which was installed during the initial construction of the home. That, of course, was the objective; a neat, finished, concealed wiring job to compliment the new lighting for the front of the home.

STREET LAMP POST

A very nice compliment to any home is an attractive lamp post at the end of the front walk. It provides light for arriving and departing guests and serves to inhibit vandalism by its illumination. Frequently, a switch can be added at the front door, or if that would be impractical, the new lamp post can be tied in with the front porch light.

Article 300 of the Code, in an exception related to residential branch circuits, permits direct burial of cable approved for the purpose at a depth of not less than 12 inches. Local inspections officials may want an "open ditch" inspection, to insure that the depth requirement is met.

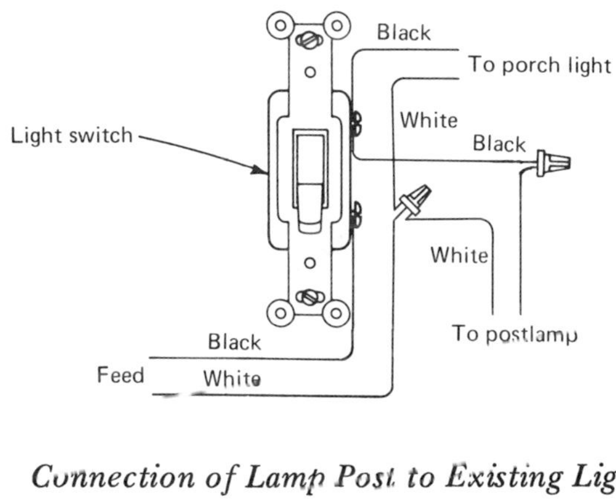

Figure 62: *Connection of Lamp Post to Existing Light Switch*

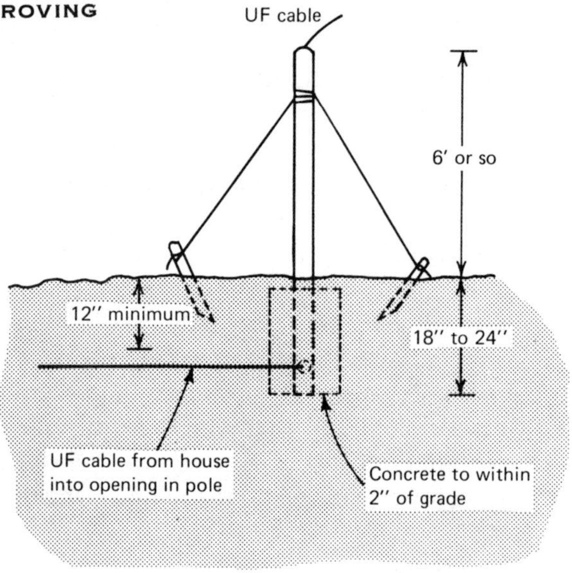

Figure 63: *Setting the Lamp Post*

The lamp post should be set in concrete after it is positioned, staked, and tied, with the cable run up into the post. Metal posts must be bonded to the grounding conductor in the underground cable. Type UF cable, rated at 15-ampere capacity, is satisfactory if the circuit into which it will be connected is 15 amperes also. 20-ampere wire should be used if the tie-in is to an existing circuit rated at 20 amperes.

In Figure 62, we show the new lamp post circuit connected to an existing porch light switch. Grounding conductors are omitted for clarity, but they would be bonded together with an approved connector and to the switch box if it is of metal.

Figure 63 depicts the lamp post set in place, with stakes and rope to hold it plumb while the concrete is poured and sets up. Lamp posts usually offer a hole for entry of the cable 3 or 4 inches from the bottom of the pole; the cable feeds through the pole and out the top, where 12 inches or so are left for connection to the lampholder portion of the lamp post, once the mix has set up.

Back at the house, the foundation will probably have to be penetrated and the cable brought into the crawl space or basement. Then, it is fished up into the wall and into the existing switch box or new switch box for connection to the switch. Be sure to de-energize the circuit that serves the switch before attempting to fish the new cable into the existing switch box. If the existing box has cable clamps, you may have to remove the switch, undo the electrical connections, and bring everything forward in order to remove the pryout and bring the new cable in under the clamp. If the box has no clamps, there will probably be a knockout in the bottom or the back of the box, and a connector will have to be installed in an inverted position to secure the cable into the box. Make sure the box will not be overcrowded; if it will be, a new switchbox is necessary.

TOOLHOUSE WIRING

Toolhouses, garages, shops, and other outbuildings situated as satellites of the main house and fed out of the main residential panelboard, may be grouped together in terms of wiring practices to be employed. Article 230 of the Code, in discussing wiring regulations for locales with more than one building or other kind of structure, permits the use of snap switches as the disconnect means, provided that the circuit extended to the outbuilding is a general-purpose branch circuit of 20 amperes or less. Even 3-way switches could be used, should it be desirable to control power to the outbuilding from more than one location.

Generally, running the wiring underground is preferable to having it overhead, as it is then out of sight and more protected against damage from falling tree limbs and other dangers that plague overhead wiring. Looking back at the section of Article 300 that pertains to minimum cover requirements, we note that as long as the branch circuit to the outbuilding is less than 30 amperes, it may be directly buried at a depth of 12 inches. A 30-ampere circuit would necessitate a minimum wire size of #10 and would call for a safety switch immediately inside the outbuilding near the point of entry of the wiring. Holding the circuit to 20 amperes would permit the use of a simple snap switch to serve as a disconnect.

But heavy equipment may be planned for in the garage or shop, in which case larger wiring would have to be run. Sizing should be based on the equipment to be served, with allowances for lighting. Circuit size requirements might be 60 or 100 amperes, and then main residential panelboard size would have to be checked with an eye toward handling the outbuilding load. Assuming a 100-ampere 230-volt 3-wire circuit to the shop, overcurrent protection of 100 amperes would have to be installed inside the house. This could be in the form of either a breaker or a safety switch that is suitable for fusing.

Now a decision needs to be made whether direct-burial cable or conduit will be placed underground to serve the outbuilding. Direct-burial cables must be placed 24 inches underground, rigid PVC must be located at a depth of 18 inches, and rigid steel conduit must be at 6 inches. Type UF cable may be selected for direct burial, but protection with conduit is necessary where it rises out of the ground along the side of a building just prior to turning into the structure. Additionally, if the circuit passes through concrete, it needs to be in conduit also, as the cable itself may not be "poured in" without mechanical protection.

Where conduit is chosen, conductors for pulling into the raceway system may be selected on the basis of Article 310 of the Code, which lists conductor insulations and their suitability. We also use the ampacity tables in Article 310 to pick a conductor size of at least #1 copper for the 100-ampere circuit to serve our shop or garage. Consult with local inspections officials to ascertain whether local requirements call for other specific methods or techniques, as these officials are the final authority in Code interpretation.

When extending service to your garage, toolhouse, or shop, it makes good sense to make the circuit 230 volts. The most difficult aspect of doing the work probably lies in getting the wiring underground, but this is more than offset by the greater flexibility that a 230-volt circuit offers, which could be very beneficial later on. For this job, you will require four conductors: two hots, one neutral, and one ground. Additional grounding at the outbuilding might be necessary also, and generally, a driven grounding rod is acceptable as the auxiliary ground.

Black and red would each be connected to one of the two hot legs within the residential panelboard. White becomes the neutral, and bare or green is the ground. Overcurrent protection would be in the main residential panelboard, and the panelboard at the outbuilding must be of the distribution type, to permit protection of branch circuits.

If six disconnects already exist in the house panelboard, adding another exceeds the maximum number permissible. Therefore, the installation of a total-service-rated main switch might be required in order to permit setting the outbuilding circuit up on a main disconnect. If the outbuilding can be served through one of the spare main disconnects, no additional main switch should be needed.

Either 100-ampere, 60-ampere, or 30-ampere outbuilding distribution panels must provide for the protection of any smaller wiring that is extended from them to within the shop or garage by branch circuit breakers or fuses. The panel could be a main-lugs-only type, with the incoming underground conductors terminated on a single 100, 60 or 30-ampere breaker, depending on the size of the underground circuit.

RENOVATING

Complete replacement of all wiring within the residence is seldom the case during remodeling, but renovation work usually entails removal and replacement of inside walls, and when this is the case, removal of older wiring becomes more practical due to its exposure as a result of removal of old wall materials. Quite often, ceilings are replaced too, and ready access is afforded to wiring previously concealed. At renovation time, most inspections officials will probably require that Code considerations relevant to new construction be applied to the structure being renovated. Such things as spacing of receptacles so that no point along a wall is more than six feet away from the receptacle become much more practical.

During renovation, our decisions in connection with complete replacement of the interior wiring system might take into account such items as whether or not the existing wiring offers a grounding conductor so that grounding-type receptacles could be installed. Knob and tube installations afford no means for grounding, as merely two conductors are run from outlet to outlet. Early cable installations frequently lack the third wire too, as does much of the BX cable in use. And while these systems are by no means unsafe from an operational standpoint, they do preclude the use of grounding-type receptacles that accept the 3-prong plug caps of many modern appliances. This fact, coupled with the accessibility of completely open walls and ceilings, would seem to mandate replacement of the wiring system.

In the same vein, insulation would undoubtedly be installed in the exterior walls before the application of new wall materials. Knob and tube, previously isolated from contact with structural members by ceramic supports and sleeves, now would be affected by insulation that might mash it against framing members. This, added to the fact that the insulation covering the wire is probably brittle, could result in exposed conductor up against wood, the very thing the knobs and tubes were to have prevented. And they did, of course, when the space they occupied was a void without insulation.

But the application of the Code requirements for new homes to one under complete renovation, and the necessity to add a significant number of receptacle outlets, would create a need for new circuits to support the additional electrical load. This is calculated by multiplying the total square footage of the residence by three watts per square foot and dividing by 115 volts to reveal how many amperes are to be handled by 15-ampere circuits. Then, as we divide 15 into the total, we arrive at the number of circuits necessary. Thus, it is discovered that an appreciable amount of new wiring must be done anyway, and complete rewiring begins to appear even more practical.

Let us, then, examine the renovation process and consider some installation situations likely to be encountered. We will start by looking at the demolition work, wherein we gain our first glimpse of the in-wall conditions with which we must contend. This differs from the more simple remodeling, where walls are generally not opened.

When exposed during the renovation process, the wiring in a forty year old home would strike us as rather skimpy, compared with wiring that can be observed in today's new home just after it is roughed in and before wall material is applied. This is because the need for outlets forty years ago was not the same as now. Nowhere near the number of appliances that we currently have was in use back then, and Code requirements in existence then that governed the installation of wiring did not impose the same regulations that they do today. Furthermore, lighting frequently consisted of one overhead fixture and one or two receptacles in each room, compared with one receptacle for every twelve linear feet of wall space in a room in a new home today.

Removal of the old wall material will usually reveal a lath onto which plaster was applied. Of course, the lath material comes down too, as in most cases drywall is reinstalled as the new finished wall. The old lath may have been wooden stripping or rock lath. Wooden strips, ¼- × 2-inch, were nailed horizontally with 3/16- to ¼-inch gaps between them to permit penetration of the wet wall material to give it something to hang onto. Rock lath was a coarse, precast material hung similarly to today's Sheetrock, but it was often covered with a finish coat of wet plaster. It was faster than wood lath, and bonded very effectively to the finish coat when it was applied. When all of this material is removed, the studding is bared, and the electrical installer's work becomes very similar to brand new work, wherein studs and joists can be conveniently bored for installation of our new plastic-jacketed cable from one room to the next.

Drywall is the material most probable to be selected for use in installation of new finished walls. A good drywall mechanic does a beautiful job in applying the finished product, with cutouts made in just the right locations to accommodate switch and receptacle outlet boxes. The electrician needs to know exactly how thick the finished wall material will be. Sometimes double-thickness drywall is installed, and this necessitates installing the rough-in boxes to protrude one inch from the edge of the stud instead of the conventional half inch.

Ceilings particularly lend themselves to installation of double-thickness drywall material, especially when the accent is on hiding nail heads and preventing their appearance at some later date when the weight of the ceiling material begins to take its toll and highlight the point where the nails were installed. This can be accomplished through installation of the first layer of ceiling material in the usual manner with drywall nails. Then, the final layer is spread with a special glue and nailed into place with special nails that feature a pulling shoulder to permit removal after the glue sets up firmly. Once the nails are removed and the nail holes filled with joint compound, no evidence can be found of where the ceiling drywall sections are held to the joists. In actuality, the strength of the double-thickness drywall is derived from bonding with the initial layer of sections, which makes use of the wide-head drywall nails to be found in general use throughout the drywall industry.

The first step in rewiring would be to remove the older wiring system, beginning with disconnection at the outlet boxes. Old cables need merely be cut away from the boxes, expediting subsequent demolition work. Heavy-duty cable shears can be used to slice away the old wiring, slicing hours off removal time in the process. All the old wiring will be disposed of anyway, so there is little incentive for meticulous undoing of the old wiring installation. Remove cable staples and pull the cables back through the holes in the wooden members. Many of the holes bored years ago for cable installation can be reused by us as new work is begun. If they employ cable clamps or offer knockouts for installation of cable clamps, some of the metal outlet boxes might also be reused. If the original installation was BX, however, the cable retainers within the outlet boxes may be unsuitable, because BX clamps engage the armor more severely than can be permitted with plastic-jacketed cable.

When the walls and ceilings are cleared, new work can begin, and it is approached just the same way we go at roughing a new home. Outlet boxes are located and placed first, to expedite routing of cable and establish a definite point-to-point plan for developing our roughing pattern. Avoid running wiring until ducts and plumbing are in place, as your installation might suffer when other tradesmen exercise their only option of removing your work if it blocks theirs. Wires are quite flexible and easily rerouted, but not so with ducts and larger drain and vent piping, frequently encountered in the vicinity of or above or below kitchens and baths. Coordination with other tradesmen before beginning is sometimes very beneficial. The location of the service equipment will probably be different from the original, and this needs to be taken into consideration when routing new cables to the panelboard location for connection after installation of the new service equipment. Our example of Chapter 1 for calculation of the residential service load would be applied as plans develop for rewiring the old home during the renovation period.

Ceiling lighting outlets, so prevalent in years past, will probably not be desired in the renovated home. Instead, a switch-controlled receptacle in rooms other than kitchen and bath may be utilized. So be sure to provide for at least one switch-controlled receptacle when doing away with ceiling lighting outlets.

While walls and ceilings are open, it may be a good time to consider how desirable new door chimes and pushes might be. The necessary low-voltage wiring can be installed now, along with the lighting circuits. Telephone wiring can be put in also, and notification of the telephone company can get permanent outlets installed at strategic locations for telephones that utilize jacks and can be moved from one jack to the other. Security wiring may be desirable as well, so that door jamb and window sash contact switches can initiate an alarm if intruders attempt entry. Remember, such an open-wall opportunity may never again exist in this home, so a well-thought-out plan needs to be devised to cover all in-wall wiring likely to be necessary to support lighting and special services desirable in the renovated home.

5

TROUBLESHOOTING

LIGHTS OUT

An obvious sign of trouble is when power is lost in all or part of the home. A complete outage, coupled with loss of power in other homes in your neighborhood, indicates that the trouble is outside of the home, and notification of the utility serving your area is in order. Such outages might occur as a result of electrical storms in the summer, or snow or sleet storms in the winter, both of which cause damage to electric power lines.

Localized outages to just three or four homes can crop up when transformers are overloaded and their protection devices open the circuit as too much power is drawn through them. Here again, this is a problem for the utility, but homeowners sometimes unknowingly contribute to the creation of this type of a problem when they connect additional equipment that constitutes a significant load without notifying the utility. Were they to notify the utility, larger transformers could be planned and installed before an overload arises.

Outages to single residences alone are infrequent, but they can be caused by falling tree limbs, which might break incoming overhead service drop conductors as they fall. But this, too, must be corrected by the utility company, since it is outside of the home. Once wiring enters from the meter, it is the responsibility of the homeowner. So problems arising within the home mean a call for an electrician rather than a call to the power company.

A complete outage on the homeowner's side of the meter is rare, but it can happen in older homes, where service equipment is small and perhaps undersized, and main fusing becomes overloaded because of irresponsible wiring that has added too many equipment circuits by bleeding off the house "mains" until their capacity was exceeded. When these react to the subsequent overload by opening up, a home can be plunged into total darkness.

An electrician responding to a trouble call of this nature would first determine that the proper voltage was available on the incoming lugs of the service equipment. Then he would check the main fusing. He would disconnect any unlawfully connected equipment, replace the main fuses, and advise the homeowner that in order to connect all the circuits, an increase in the size of service equipment would have to be made.

Partial outages occur when just one of the two main fuses opens as a result of overload. Since 115-volt loads are connected line-to-neutral, loss of power might occur when the more heavily loaded main fuse burns out and all of the 115-volt outlets that are tied to its side of the 115/230-volt service are isolated. A volt-ohmmeter can check up on the continuity of each of the main fuses to find the offender, and replacement of the fuse can be made after disconnecting some of the overloading equipment. Here again, an update in service equipment would appear to be in order to permit complete flexibility in the use of equipment within the home.

Homes that experience continuing difficulty with one of the main fuses consistently overheating and burning out might benefit from some rearrangement of the load within the existing panelboard. Through careful study of the household's operating characteristics, it may well be determined that transposing one or two heavily loaded circuits on one leg of the incoming service with one or two lightly loaded circuits on the other leg will produce better balance loading across the two legs to neutral, with respect to 115-volt circuitry. It is not rare to find one leg serving most of the lower-voltage load within the residence in cases where smaller services are involved, and realignment can ease the demand for current in one of the service entrance conductors by shifting some of the lower-voltage loading to the other conductor. This suggestion is not to be taken as an alternative to a service change when one is needed. Instead, it is merely a remedial step to be employed in a home where the service is adequate, but better balancing appears necessary to facilitate cooler operation of main fusing.

Oftentimes, 100-ampere services are candidates for rebalancing when clothes washers, dishwashers, kitchen circuit, a predominant portion of the lighting, and small window unit air conditioners might be branch-circuited from one side of the panelboard's bussing. Evening mealtime in the summer and the hours immediately thereafter might evolve into the time when trouble begins to develop with main fusing on just one side of the panelboard, perhaps pointing up the need for considering better balance-connecting.

By far the most common type of outage within the home occurs when the overcurrent protective device, either fuse or circuit breaker, of a particular branch circuit opens the circuit on overload. When this happens, it is important to detect the cause of the outage; merely replacing the fuse or resetting the breaker is not enough. Take remedial action against the cause of the problem, as well as the effects. Let s identify some of the possible causes.

Electric hand-held hair dryers are enjoying great popularity today, as Americans take advantage of the time saved through electrically blow-drying their hair. But a 1,000-watt dryer loads a 15-ampere circuit to 58% of capacity, and on circuits that are already partially loaded, this may be enough to cause reaction of the branch circuit's overcurrent protective device, fuse or breaker. When Type S fuses are involved, replacement with one of the same rating is mandatory; otherwise the circuit will not be restored. Homeowners frequently experience difficulty in getting the outage rectified when they unknowingly select a replacement Type S fuse that is of a different rating than the original, and it will not make contact because of its special design which makes it compatible only with an adapter with the same rating as the fuse. Unbeknownst to some homeowners, they are purposely designed this way to prevent overfusing. But replacement of the fuse constitutes action against the effects only; reconnection of the hair dryer to a heavier circuit would be the correct type of action against the cause of the problem.

As summer comes, air conditioning begins to constitute a significant electrical load within many homes. Window units may be purchased and installed, and 115-volt models may be plugged into any convenient receptacle, oftentimes with problems arising. Article 440 of the Code, under "Branch Circuit Requirements," reminds us that the rating of a window unit may not exceed 50% of the branch circuit's capabilities when other lighting or appliance outlets are served by the same circuit. But this precaution might be overlooked, and then intermittent blowing of the fuse or tripping of the breaker may crop up. The correct answer here is to circuit the air conditioner separately and size the circuit so that it is at least 125% of the nameplate rating of the unit.

Underground cables that have been in service for many years can create difficulties if their insulation and jacket begin to break down, permitting leakage of current to ground. In such a condition, branch circuits will go dead as the protective device opens up to prevent short-circuit current flow to earth. This particular type of problem may be first observed during wet weather, when the ground is more conductive; and the problem may appear to go away during dry weather. At first, the occupant may be baffled when his living room circuit goes out, because invariably, pole lights out front are switched from just within the front door and tied into the branch circuit that serves the living room and perhaps other rooms in the home. But if he switches the pole light off and then restores the circuit, operation will continue, because the underground line is isolated from the branch circuit hot leg by virtue of the open switch. The answer to this problem is replacement of the underground cable with new cable of the same size and material as the conductor of the circuit into which it is being connected. The cable, which should be Type UF "Underground Feeder" cable, is in a protective jacket approved for the purpose.

And here's a problem that can arise where ground-fault interrupter protective devices are employed to protect outside receptacles. If the weatherproof cover or gasket does not keep out all water, even in heavy rains, the buildup of trapped water within will set up a short circuit and allow enough current to trickle to ground to cause GFI protectors to react. If other outlets within the home are on the same circuit as the outside receptacle, portions of the home will be out of electricity also. And a GFI breaker will not reset, and this can be very baffling to the unsuspecting. The solution here would be the evacuation of all water from within the outside receptacle and replacement of the weatherproof covering with one of good integrity.

When circuits go out, understanding and judgment are to be employed in a detective-like fashion until the difficulty is discovered. Then, repairs or replacement, consistent with Code recommendations, should be instituted to clear the problem and restore service.

TAMPERPROOF FUSING

There is serious danger in overfusing. To combat this, the Code in Article 240, Division E, requires plug-type fuses to be of the Type S tamperproof variety and limits the use of standard Edison-base variety fuses to existing installations where there is no evidence of tampering or overfusing and their use constitutes replacement only.

The Type S fuse utilizes an adapter that is first installed into the screw-shell plug fuse holder. Three sizes of adapters are readily available based on current ratings to fit 15-, 20-, and 30-ampere Type S fuses. If a Type S fuse of a rating larger than that of the adapter is installed, the circuit will not be restored, because dimensional differences between the adapter and the incorrect fuse prevent completion of the circuit. The unsuspecting will oftentimes attempt replacement of a blown Type S fuse with one of a higher rating, and not knowing why service remains out in the affected area, they will frequently be unsuccessful in this attempt and may wind up having to call for an electrician. This is a situation that frequently develops, and it is not at all unusual for an electrician to discover on his service call that Type S fuses of the wrong size are installed in unmatching adapters. Therefore, it is important that individuals that are involved in Type S fuse replacement insert a new one of the same type and rating as the one being replaced.

A blown fuse points up a need for investigation as to why the fuse blew in the first place. However, a fuse that has been in service for several years on a partially loaded circuit may fatigue and eventually break down, and in such a case mere replacement may be all that is needed.

A distinction between Edison-base plug fuses and Type S plug fuses may be made by comparing the threads. An ordinary Edison-base plug fuse has a copper, light-bulb size thread, whereas the tamperproof Type S plug fuse exhibits ceramic threads that are finer, and smaller in cross-sectional area. The two may not be interchanged. 15-, 20-, and 30-ampere Edison-base plug fuses will all fit in the same socket, but not so with Type S fuses, which, by design, cannot be installed as direct replacement for each other. For maximum protection, in your own home or in rental property which you may have or hold for others, Type S fuses should be installed, correctly selected and sized to protect branch circuit wiring within its capacity. It is a good investment to have an electrician make this installation for you, if yours is a fusible panelboard with standard Edison-base plug fuses. Under no circumstances should you attempt to defeat the tamperproof adapter. If you are continually blowing fuses, an additional circuit to relieve some of the electrical load is the correct answer. Remember, proper fusing in the electrical panelboard represents protection for you and your property. Be sure branch circuits are properly fused to match the size of circuit conductors. If in doubt, have the fuses checked.

GROUND LOSS

Once in a great while, the homeowner may justifiably feel that his electric bill is excessive. There have been cases where electrical losses to ground within the premises of the customer's wiring system have resulted in his meter registering the flow of more power than he actually used. The fault lies not with the utility, but with a high-resistance short to ground either in the wiring within the home or in an outside branch circuit served by the house power panel. The power company can check the accuracy of their meter, but if current is actually flowing and their meter is accurately registering the flow, Mr. Homeowner must take the measures necessary to determine if the usage within his home is intentional and beneficial, or if there is a problem of which he has been unaware.

A high-resistance short is one in which the fault-current flow is less than that required to trip the breaker or blow the fuse. A ground-fault type of circuit interrupter could distinguish between proper usage and ground-fault flow and would open the circuit. But ordinary breakers and fuses do not incorporate the comparative kind of network that recognizes the difference between fault-current being lost to ground and legitimate current flow properly occurring over hot and neutral.

You yourself can easily check interior wiring for fault-current flow by de-energizing every load within your home. Turn off all lights, pull the disconnect on the furnace or turn the thermostat all the way down (when in the heating mode), unplug electric clocks and other automatic loads, such as refrigerator and freezers. The range may include an electric clock, so pull its breaker or remove its fuses. Or better yet, unplug the range, so that its branch circuit is subjected to the fault-current inspection, also. If you have any electric baseboard heaters, make sure their thermostats are off. Then, take a look at your electric meter. It should not be registering any current flow. You should see a still meter. If not, disconnect all branch circuits one at a time, and examine the meter for effect each time one is disconnected. When you find the branch circuit which when isolated stills the meter, disconnect the hot conductor from the power panel terminal, and, using an ohmmeter, check for continuity of circuit between the disconnected conductor and ground. If you see a completed circuit, the conductor is guilty of fault current flow. With an *accurate* ohmmeter that features multiple resistance ranges, you can discover the quality of ground that exists. This answer will be the number of ohms that the resistance has. By dividing 115 volts by the found resistance, we determine amperes. Then, multiplying amperes by

115 volts, we come up with the amount of watts that the ground loss consumes. This times 24 hours in a day, divided by 1,000, gives us kilowatt hours per day, and this times the kilowatt-hour rate tells us the additional cost to us due to the ground loss. In making this test, be sure all loads are identified and properly disconnected. You can have closet and attic or basement lights on behind closed doors and not realize that they are burning. Make a careful check throughout the home to be certain everything is off.

Suppose that the ohmmeter shows a circuit to ground on the energized conductor with a reading of 30 ohms. Dividing 115 volts by 30 ohms produces 3.8 amperes, and 3.8 amperes multiplied by 115 volts equals 437 watts, the equivalent of four and one third 100-watt light bulbs. Multiplying 437 watts by 24 hours a day yields 10,488 watts over a 24-hour period. Divide by 1,000 and the answer is converted to 10.4 kilowatt-hours, the unit of measurement that the utility bills by. This ground loss approximates 10½ kilowatt-hours per day. If the utility charges 6¢ per kilowatt-hour, it's costing 63¢ per day. This times 30 days yields $18.90 per month.

The next step is to locate the ground loss and correct it. If you have an underground yard-light cable on the offending circuit, and if the controlling switch is incorrectly wired so that it breaks the neutral instead of the hot, a deteriorating underground cable could leak power to earth ground. Breaking the joints to the underground cable in the switch box and rereading the offending circuit for grounding might reveal that to be the problem. The solution is to replace the underground cable and reconnect the switch to break the proper leg.

Sometimes, sharp bends made in BX in armored cable installations will mash conductor insulation. Over a period of time, deterioration of insulator quality will permit a high resistance circuit to be made between the energized conductor and metal jacket in a stress situation, allowing fault current to flow. Working from the outer end of the circuit with respect to the panel, open joints in the wiring one at a time and reread the line for the grounded condition. Eventually, you will identify that segment of cable wherein the fault lies, and replacement of that piece can be made.

A nail driven into a cable in such manner that a ground was almost, but not completely, established between energized and grounding conductor, may, over a period of time, become a poor enough conductor to set up a small current flow between the energized conductor and its damaged insulation and the grounding conductor. The same isolation techniques will identify where the trouble is and permit replacement measures to be taken.

A positive, definite short will activate protective devices. But a marginal situation, such as the one we have just discussed, may not create the clean short-circuit path necessary to trip the circuit protector. Instead, current flow occurs at a low enough level to go undetected by our branch circuit overcurrent protective device. And the meter registers the flow, just the same as if we had left the lights on. These situations are rare, but if you have real reason to suspect ground loss, contacting an electrical contractor is perhaps in order. He can send the qualified personnel to make the necessary checks to determine if your problem is real. And the utility serving your area will, in all probability, be happy to make a validity check on their meter. After all, they're interested in your paying only for what you really use, too.

OPEN NEUTRAL ON A MULTI-WIRE CIRCUIT

The advantages of using multi-wire circuits have been mentioned in earlier sections. A common neutral can handle two circuits if the energized conductors are connected out of phase with each other. This is because the two 115-volt sources offset one another when brought together at the neutral point, with the common neutral maintaining balanced voltages in each of the two circuits by grounding out any imbalance that might occur as a result of varying loads.

If the common neutral is lost for any reason, voltages will be unbalanced in each of the two circuits, and lighting served by them will exemplify this by a distinct change in brilliance. This can sometimes happen as the common neutral pulls away from its joint with the older neutrals of the two separate circuits and leaves them still in contact with each other.

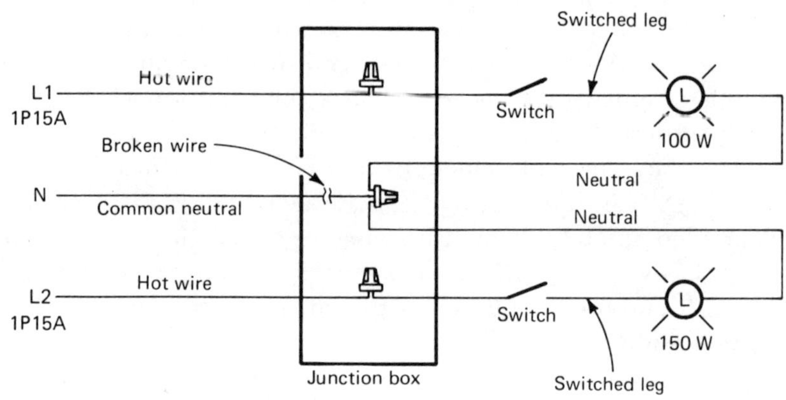

Figure 64: *Loss of Common Neutral on Multi-wire Circuit*

In such a situation, the imbalance set up in the two circuits will be obvious to Mr. Homeowner, as he observes the reaction of light bulbs in his fixtures. Some lights will be brighter, but some will be dimmer. Junction and outlet boxes will have to be examined to find the opened joint, so that restoration can be made. We need to take a look at the reason for this situation. Figure 64 shows two light bulbs on two different circuits. One is 100 watts, and the other is 150 watts, and their respective circuits are served by a multi-wire circuit from the panelboard.

A break in the neutral joint in the junction box that isolates the connected neutrals of the two circuits from their bond with the common neutral has the effect of impressing 230 volts across the 100-watt and the 150-watt bulbs, in series. The two 115-volt sources are connected L-1 and L-2 in the panel, so the effective voltage in the remaining circuits that are in series with one another doubles. But the two bulbs offer different resistance to current flow, with the greater resistance being in the smaller bulb. So the voltage divides unevenly across the two, and they burn at different brilliances. The 100-watt bulb becomes brighter, as approximately 147 volts appears across its filament, while the 150-watt bulb becomes dimmer, as approximately 83 volts appears across its filament. With the common neutral restored, our two circuits are rebalanced, and 115 volts now appears across both bulbs.

To grasp the real significance of this example perhaps requires further discussion. Alternating current is reversing its direction every 1/60 of a second. Thus, in the course of one second, 60 reversals occur, each of which is referred to as a cycle. Plotted against a straight line, the result is the familiar sine wave, where magnetism rises to one side and then drops to an equidistant position on the other side of the zero base line before beginning its wave form movement to the other side again. At the rate of 60 cycles per second, this reversal of direction is unobservable in residential lighting because it happens so fast. But the reversal creates the alternations that set up the attraction of alternating current to ground and to out-of-phase legs within the same circuit. When two legs of a polyphase circuit are referenced to one another, full circuit voltage is impressed across the load. In 3-phase circuits when two loads are connected in series between two out-of-phase legs, and a grounded conductor is connected between the two loads, voltage is reduced across each load by 57.736%, the reciprocal of which is 1.732. And current flow is between each leg of the circuit and ground. Since the ground is present to ground out its side of the circuit, different current flows can occur in the two loads, with full reduced voltage being maintained. We are talking about 3-phase power, because this is the way power starts out and is delivered through the utility's distribution system. These three phases,

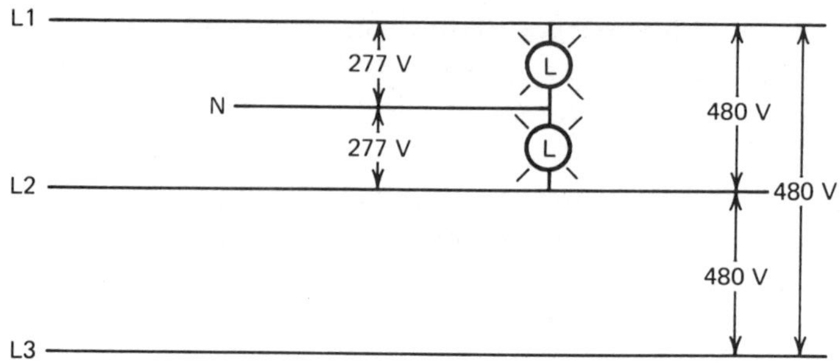

Figure 65: *Single-phase Power Tapped from 480-volt 3-phase Circuit*

each 120° out of phase with the other, compose a complete 3-phase circuit, with each leg exhibiting a distinctive sine wave. The 360° rotation of the generating station's rotor through three magnetic fields within the generator produces three separate phases that lend themselves to use in commercial, industrial, and residential applications in a variety of ways. Although our considerations in this text have to do with single-phase residential power, we need to know how this single-phase power is wrought from the basic 3-phase circuit.

Figure 65 illustrates a commercial 3-phase circuit in which derivation of single-phase power is shown for the 277-volt fluorescent lighting load represented by each of the small circles with an L in the center. Power legs L1, L2, and L3 have a potential to each other of 480 volts, but only 277 volts between either of them and the neutral represented by "N." Dividing 480 volts by 1.732, we arrive at 277 volts. Thus, the lower figure of 277 volts can be used for lighting throughout a commercial building, and frequently is. The important thing here is recognition of the fact that each of the power legs offers a potential to ground, albeit a different one from that which exists between any two of them. In today's commercial buildings, 480-volt 3-phase circuitry is quite prevalent.

Suffice it to say that utilization of a neutral in a 3-phase circuit establishes a new voltage between the neutral and any of the three power conductors. The neutral can have loads connected between itself and each of the 3-phase legs at different points throughout the electrical system, and it may serve all three of the phases, because loads so connected tend to cancel the loading effects of one another at the neutral point. This is typical of a multiwire circuit. The neutral serves as a common conductor back to the panel for a 4-wire circuit shown in Fig. 65. Flow of current to it from any other leg tends to reduce its load, not increase it, as the 120° out-of-phase relationship between the power conductors sets up an attraction between different leg loads that nullifies a portion of the potential that is referenced to ground. This important characteristic gives rise to the multi-wire circuit and to the flexibility that is achieved through staggered single-phase connections to the 3-phase circuit of Figure 65. In essence, we have three individual single-phase lower-voltage circuits with just four wires.

Typically, a service lateral to a segment of a residential service area will consist of one high-voltage phase conductor, plus the ground wire. Voltage reduction is developed through the use of a center-tapped single-phase transformer that produces the familiar household voltage from a much higher voltage. The lower-voltage windings have voltage potential rippling through them that is the same as the sine wave exhibited in the higher. Then, by grounding the transformer's secondary midpoint, we establish the typical, grounded, residential, single-phase service directed into the home via the service drop.

Figure 66 depicts a typical overhead service drop as it might be connected into a single-phase high-voltage service lateral. There is an electrical

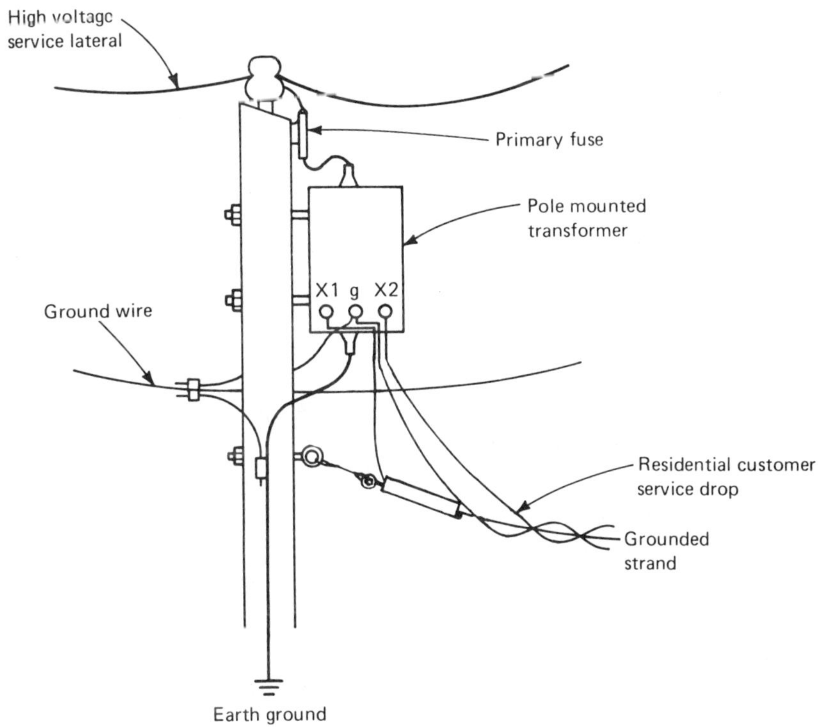

Figure 66: *Pole-mounted Residential Service Transformer*

voltage of 230 volts between X1 and X2, and 115 volts between either X1 or X2 and ground, designated by "g." So the capabilities for our residential 3-wire multi-wire circuit actually originate at the transformer atop the electrical service pole serving the residential area in question. For all practical purposes, we can say the 230-volt single-phase circuit on the secondary side of the transformer is divided and grounded in the middle to give us the dual voltage potential within the home of both 115 and 230 volts. Lighting can be served with the lower voltage, heavier loads with the higher voltage. The "g" conductor is grounded both at the pole and in the house service equipment.

The residential multi-wire circuit has all three conductors of the service drop duplicated in branch-circuit fashion within the multi-conductor cable that is extended to serve outlets that will be tapped from the multi-wire branch circuit. At some point, each energized leg is impressed across a load, the other side of which is connected to an identified neutral conductor that makes its way back to the common neutral of the multi-wire circuit. Our common neutral will not be subjected to a load any heavier than the load handled by the most heavily loaded of the energized conductors. And even then, a portion of its attraction to ground over the neutral wire is offset by the total wattage in the other leg due to the "out-of-sync" relationship between them. The neutral grounds out *only* the imbalance remaining.

Opening of the common neutral in a multi-wire circuit removes the ground reference for each of the two out-of-sync energized wires, so that they "see" each other through their connected loads. When this happens, the resultant 230 volts splits unevenly, depending on the resistance imposed on either side of the now missing neutral, and unequal voltages appear across connected loads. Hence, with reference to Figure 64 again, the resultant 147 volts and 83 volts mentioned in connection with it were the products of unevenly dividing voltage when the neutral was lost. You may observe this yourself one day in your home or someone else's where multi-wire circuiting was employed, in the event an occasion arises wherein the common neutral is opened for some reason. Such an occasion is, however, infrequent, and even then is limited to homes where multi-wire circuiting has been employed. And multi-wire circuiting is an accepted practice, as far as the Code is concerned. Properly done, with tight joints, difficulties should never arise. But being able to recognize the malady when it does occur is important.

The ordinary colors of insulation that are encountered in 3-conductor cable used for residential multi-wire circuiting are black, red, and white, with white being the identified, grounded conductor, or neutral. The energized conductors, red and black, can be observed within the panelboard connected to branch circuit overcurrent protective devices, each fed from an opposing 115-volt power source. In the event that rearrangement of breakers within a panelboard becomes necessary for some reason, it is *imperative* that the 230-volt relationship of the two energized conductors of the multi-wire circuit be maintained. If not, the common neutral will be subject to overload, because 115-volt circuits derived from the *same* side of the panelboard do not offset one another at the neutral point; and without this offsetting characteristic, the current flow in each circuit must now be carried to ground by the common neutral. So if the two circuits in question were to be partially loaded to 9 amperes each, we would find 18 amperes appearing in the common neutral. But in the properly connected multi-wire circuit, the common neutral would never have to carry more current than the heavier-loaded of the two energized conductors.

GLOWING LAMP POST

When first experienced, it's a strange phenomenon. The lamp atop the post in the front yard appears to be trying to come on by itself—even though the switch that controls it is off. This occurence is most likely in wet weather, particularly when the ground is very saturated, as it is during a rainy spell. In dry weather, the problem may not be apparent. You wonder as you watch the bulb glow within the post fixture. How can that be? How can the light come on at all when the switch is in the off position?

Two conditions are sometimes combined that cause this difficulty. Proper wiring of the switch that controls the lamp post calls for the energized conductor to be broken via the contacts of the switch, but occasionally it is discovered that the switch may have been installed by someone other than a professional electrician. When this is so, it may be found that the neutral conductor of the cable serving the lamp post is actually connected through the switch, while the energized conductor, which forms the other half of the circuit, is intact all the way to the fixture outside. Secondly, deterioration of the underground cable may have taken place, either because of age, or through accidental damage caused by someone digging without due concern for where the cable was buried. This deterioration would lead to breakdown of the cable jacket and insulation and is most prevalent where cable unsuitable for the purpose was buried underground. The conductors, because they would no longer be satisfactorily insulated, would allow current leakage to ground, particularly when the earth is moist and conductive.

Remember, our residential wiring system is a grounded single-phase system, where reference of the power conductors to ground across the load served initiates current flow. When improper switch wiring is combined with deteriorating insulation of the underground cable in rainy weather, we may observe glowing of the lamp post when the switch is off.

In considering the installation, we might discover that a circuit has been established for current flow through the lamp post light bulb out to earth ground. The cable may have been tied in to any convenient branch circuit in the crawl space under the house with the power off. Then this cable might have been extended up into the wall to await installation of a switch above. For someone who didn't appreciate the significance of the white and black insulation on the wiring, it might have been easy to select the wrong wire for connection through the switch. In such a case, the switch would only open the returning neutral side of the lamp post circuit, which would leave the other conductor that runs out to the fixture and back to the switch energized at all times. Ordinarily, the switch would control the lamp satisfactorily, because when it would be opened, the grounded neutral connection would be removed, isolating the circuit through the bulb from ground. But if current leakage occurred underground from the neutral wire whose insulation was no longer effective, current would flow at varying levels to this "new" grounding source for the neutral, depending on the conductivity of the earth. If the weather were wet, this would be relatively high.

Reconnection of the switch so that it breaks the hot wire to the lamp post is in order, along with replacement of the underground cable. Replacement of the underground cable may be accomplished without removal of the old one. Disconnect the old one and abandon it in place, cutting off cable that is above ground in the crawl space down to a level just below grade to insure complete burial. A check with local inspections officials may be in order, here, as some municipalities may insist on removal of underground cables so as to insure the impossibility of reconnection. Then, a new ditch can be dug to a minimum of 12 inches, and replacement can be made with type UF underground feeder cable as approved by the Code. Be sure that the underground wiring complements the circuit into which it is tied in type and size of conductors.

Penetration of house foundation below grade is preferable to above-grade penetration, with wiring turning up inside the crawl space, but whether it turns up outside or under the house, use conduit from below grade up beside the foundation. Outside, use a conduit body to effect the 90° turn to enter the crawl space. This is necessary for mechanical protection of the UF cable wherever it rises up from below ground. Most often, NM sheathed cable, as used generally in interior wiring, will be extended into the crawl space into a junction box, and underground UF cable will be jointed at that point and continued on outside underground. UF is more expensive, so minimizing its use can help save some on the installation.

230 VOLTS IN THE SWITCH BOX

Here's an incident that can be quite alarming to the unaware. Should it become necessary to replace a switch or extend a circuit from a multi-gang switch box in which a 3-way switch is located along with another switch, it will sometimes be found that the voltage across two conductors will be 230 volts. This discovery could be upsetting to an amateur who randomly disconnected both switch wires and is attempting to get everything back together again. He may decide to turn the power back on and see which wires are hot, using a meter or test light. When he does, he may discover a potential difference of 230 volts across two particular wires. On occasion, this has been very disconcerting to a few people.

Upon reflection concerning the practices relating to 3-way switch wiring, we find that this is not an oddity. Consider a downstairs hall light controlled from both upstairs and downstairs. The light itself is on a downstairs general lighting circuit, and its switches form a part of that circuit, even though one is upstairs, perhaps in a double gang switch box. Another switch, controlling the upstairs hall light, is on a different circuit altogether from the downstairs hall light. And these two lighting circuits could very well be connected to overcurrent devices fed from different legs back at the panelboard. Remember, the circuits can be fed from either of the two hot service conductors in the panelboard, depending upon how they were connected.

So, of course, if you select the two conductors in the switch box that are parts of separate circuits that are connected in opposition to each other, you will read 230 volts, and that's not uncommon. But don't be alarmed. Mere replacement of a switch calls for nothing more than reconnection of the new one in just the same way the old one is connected. Removal of both switches and disconnection of all the wiring is unnecessary. If enough slack is left by the electrician, the switches can be brought forward, permitting replacement without opening of splices inside.

The obvious caution is that both circuits should be de-energized. If you have the option of doing the work in the daytime, opening the main service disconnects will isolate everything in the house and assure you of absolute outage for safe work. If you must work at night, and you need light, an extension cord of the heavy-duty variety could be plugged into a special purpose receptacle, such as a washer or kitchen counter receptacle, and all general lighting circuits could be shut down by removing fuses or opening circuit breakers. Then, by using a drop light, you could have light where you need it and at the same time be assured that the circuits on which you're working are dead. In either case, day or night, the existence of 230 volts between the two circuits need not concern you if your work is being correctly performed.

REPLACING OR INSTALLING DIMMERS

In addition to special fixtures designed for accent lighting, special effects can be created through the use of dimmers installed to control existing fixtures, often enhancing the beauty of the existing lighting many times over. Readily available from most electrical supply sources, dimmers are easy to install and provide tasteful moderation of light to suit the occasion for a relatively small expenditure. In one-switch or multi-switch installations, they merely replace one 2-position type of switch and effect softening of light output over almost an infinite range.

Special care is required during installation or replacement of a dimmer. Practically every manufacturer packs instructions with the unit that insist upon de-energization of the circuit serving the switch and light, and for good reason. Sudden spikes of current through the unit can destroy its ability to modulate power to the light fixture. Trying to install a fixture on an energized circuit, in addition to being contrary to good practice, represents a hazard to the control unit portion of the dimmer itself. Power is not to be restored until the dimmer is permanently in place, properly wired, and ready to be placed into use. During installation, the circuit on which the dimmer is located *must* be isolated. This rule is probably known by most people and is generally observed. However, this leads to another principle relating to lamp replacement in a fixture that is controlled by a dimmer: when replacing burned-out lamps, be sure the fixture is turned off. Replacement of lamps with the fixture still energized can also be destructive to the dimmer. Though many don't realize it, removal of the old lamp from its screw shell sometimes results in momentary contact between screw-shell fixture wires if the socket attempts to turn when rotary removal motion is applied to the old bulb. Momentary shorting of the socket leads with the power on will destroy the dimmer, as its maximum rating is most often in the range of 600 watts for ordinary residential use. Around five amperes is all they can stand, and peaks above this level wipe out dimming control capability. Even if the socket is firmly anchored, the smaller candelabra bulbs sometimes separate at the point where the glass meets the base when being turned, and shorting of the filament leads usually results. This has happened, and when it does, owners wonder why the dimmer will no longer function properly even after the old base has been removed with the power off and the new bulb has been installed.

The option in dimmer selection is between 2-wire and 3-wire control. If only one switch is to control the light, a simple 2-wire model is sufficient. But if two or more switches will control the light, a 3-wire model is required. Recall, from the discussion in Chapter 1, that the 3-way switch is single-pole, double-throw, allowing alternate paths for the light circuit from one switch location to the other. So it is with dimmers. Each one incorporates an on-off switch in addition to, and in series with, the current regulation module. But the switch in the 2-wire model is single-pole, single-throw, whereas the switch in the 3-wire model is single-pole, double-throw. The dimmer merely provides power modulation in series with on-off control. And the modulation module is the limiting factor in regard to current-carrying capacity, which is the reason why extra care is necessary with respect to turning off the power when installing or replacing dimmers.

In installations where three or more switches control common lighting outlets, it will be necessary to find one of the 3-way switches and make our replacement there. If in doubt, remove the switch trim plates, and examine the terminations to each with a service light or flashlight. Or, if necessary, remove the switch mounting screws with the power off, and bring the switch forward enough to permit examination. You will have two choices of dimmer location in multi-switch applications, because there will be only two 3-way switches, the balance being made up of 4-way switches.

Careful installation is next, followed by remounting of the dimmer. Manufactures currently offer more smaller-body dimmers than in previous years, so no difficulty with box space should arise, unless the box was overcrowded to begin with. With power restored, you will enjoy the special effects of your new dimmer for many years to come.

REPLACEMENT OF LAMP SWITCHES

The average modern home has quite a few lamps throughout the house. Their number may have been increased over the years through the use of the switch-controlled receptacle outlet thereby eliminating the need for overhead ceiling fixtures. Living rooms, dens, bedrooms, libraries, studies, and offices in the home call for decorative and functional lighting, often accomplished with table lamps and floor lamps. Many styles are available, with different variations in light diffusion to suit the area served. Switches in each lamp permit local control to select on-off, 3-level, or infinite control over the range of the bulb installed, depending on the type of switch used. From time to time, these lamp switches require replacement, as they wear out over the years from use. Let's investigate switch replacement as we examine three types of lamp switches frequently found in use. They are the rotary switch, pull-chain switch, and push-through-shell lampholder types. Quite often, they are fitted at the bottom with an 1/8-inch female threaded collar to permit mounting on an 1/8-inch lamp nipple. Such a nipple may run the length of the lamp, serving to hold the sections together as well as providing an accessway for the lamp cord.

Each switch is really a bulb-holder-and-switch combination, encased in a copper, removable jacket, with insulator and opening for switch actuator mechanism. When trouble develops with the switch, e.g., the lamp won't turn on or off, or you have to jiggle the switch to get it to work, replacement is in order. Internal components that establish electrical contact do wear out, and for this reason, replacements are readily available from most electrical wholesalers and many hardware and variety stores.

First, *unplug* the lamp! Then, remove the lampshade and bulb. The switch jacket is snap-fit into a stationary lower rim, which is notched inwardly around its circumference to form holding cleats that grasp the removable jacket when it is pushed into place. Depressing the jacket with your thumb beside the opening for the knob releases the rim's hold on the near side. Tilt the jacket outward, and it will then come free and pull off, exposing the replaceable shell-lampholder type of switch within. The stranded lamp cord that is connected to the two screw terminals must be disconnected. If there is not enough slack for you to work the wire easily, you can probably pull some wire through the nipple raceway if you encourage it to enter the cord opening at the bottom of the lamp while you tug gently at the switch end.

Once the switch interior is free, you can install a new interior. Then, reverse the above steps for reinstallation of the switch and protective jacket. Be sure to preserve and reuse the cardboard insulating sleeve found inside between the jacket and switch interior. This is *very* important as a guard against contact between the energized portions of the switch lampholder and the protective outer jacket. It is preferable to have the energized conductor of the lamp cord terminate on the screw terminal that connects to the button in the base of the bulb holder, with the grounded neutral conductor of the lamp cord terminating on the other screw terminal, which connects to the screw-shell portion of the bulbholder. In this manner, accidental contact between the energized portion of the bulb holder and the lamp proper is minimized. However, with simple two-prong plugs on the end of the lamp cord that permit insertion into the receptacle either way, assuring such maintained polarization within the switch is impossible.

If the plug is of the polarized or grounding type, then care can be exercised to insure connection of the energized prong of the plug to the "button" terminal of the switch.

The simple push-through, slide-bar type of lamp switch makes and breaks the electrical circuit through it as the slide-bar is pushed through from one side to the other. Pull-chain lamp switches feature a spring-loaded retraction mechanism that repositions the chain-actuated latch to rotate revolving switch contacts within for a break-make, off-on sequence each time the chain is pulled. Rotary switches utilize knobs that are twisted in such manner as to operate internal cams or cogs for opening and closing tension-held contact points. You can oftentimes go from one type of switch to the other if a change in the method of turning the lamp on or off is desired. But the slide-bar variety requires a dual-opening cover and insulator, and some variation may be noted from one type of switch to the other in regard to the type of jacket and the number and size of openings required. So for changing the type of switch, it might be advisable to purchase a new shell-lampholder-switch assembly, complete with jacket, insulator, and threated base rim.

Four-position rotary switches are needed in 3-level lamp applications. The bulbs for these lamps offer three levels of lighting through two filaments, which are alternated or burned simultaneously to effect the three levels of brightness. These lamp switches are called 3-way lamp switches and make contact sequentially when turned through their full range to light the smaller filament first, then extinguish it and light the larger filament, then hold the larger one on while relighting the smaller filament for full output. Special contact points in the base of the bulb holder mate with corresponding contact points on 3-way bulbs. Ordinary bulbs may be used in some 3-way lamp-switch bulb holders, but two positions of the switch will be ineffective. With an ordinary bulb installed in a 3-way switch, the sequence would probably be nothing-on-nothing-off; that is, each click in between on and off causes no change in lamp status.

For best results in replacing the lamp switch, remove the old switch and take it with you when purchasing a replacement. Then the new switch can be compared with it for size, action, style, and fit. After the new switch is installed, pull the lamp cord back out of the base to remove unnecessary slack and to induce the new switch interior to nestle in the fixed rim. Then, replacement of the insulator and exterior jacket will not crowd or pinch any cord within the switch housing. And in terminating lamp cord, start with freshly stripped wire ends if conductor fatigue is evident when the old switch is removed. Strip away just enough insulation from the wire to encircle the screw terminal, and twist the exposed wire to prevent it from spreading when tightened. Tinning the tips with solder assists in maintaining the twists.

An underwriter's knot should be formed in the separated cord leads before connecting the bared ends to the screw terminals. This knot will lie in the base of the lamp switch in order to prevent the strain that is created in the cord when it is pulled from being imparted to the terminated connections. Thus, if the lamp is pulled or the cord is accidentally snagged, the sudden tightening and pull of the cord will not draw loose an energized conductor and set up a potentially hazardous situation within the lamp whereby the conductor would come in contact with the exposed metal parts of the lamp. The underwriter's knot is formed by inserting each lead through a loop formed by the other lead. Note in Figure 67 how one lead of the separated lamp cord is woven through a loop formed by the other lead, after first forming its own loop. One lead is looped toward the front, and the other toward the back, and then the knot is formed by passing each through the other's loop. The knot should be drawn snug, but not tight. Once terminated on the screw terminals, it cannot come undone, but remains in place to guard against too much tension being communicated to the terminations.

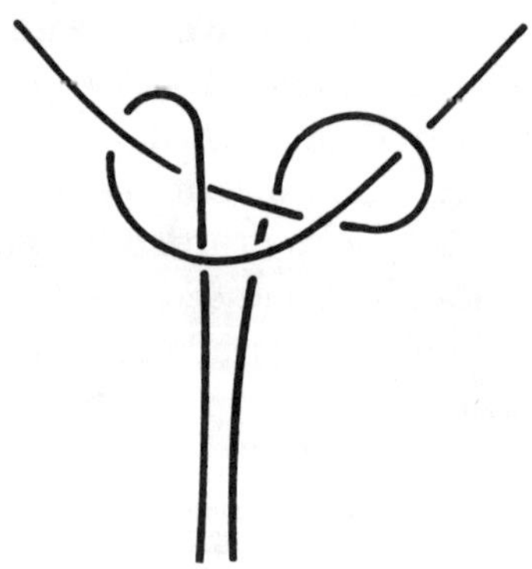

Figure 67: *Underwriter's Knot*

When the switch is replaced, it may not be necessary to disturb the knot; merely pull it outward enough to gain slack so that the change can be made. Then, retying the knot will not be necessary.

LOOSE SWITCH OR RECEPTACLE

Outlet boxes are supposed to be installed so that edges of the box are flush with the finished wall. A ¼-inch setback is permitted, however, in walls of noncombustible material. In such cases, the ears of the device, be it switch or receptacle, pull down flat on the finished wall. But if the wall material is displaced in the area where the mounting ears of the device make contact, the switch or receptacle will be loose, as there is no support. Were you to remove the breakaway ears and try to mount the device hard to the box itself, it would be recessed into the wall too far to permit proper protrusion through the trim plate. This problem can be encountered in older homes, where wall material might have broken away over the years, or even in newer homes, if sufficient attention was not given to maintaining close tolerances between box and abutting finished construction.

Sleeving the mounting screws will permit firm anchoring to the box itself in such a case. Measure the distance between box and finished wall. Using ¼-inch OD copper tubing, cut a sleeve of the same length and slip it over the mounting screw just before you begin to thread it into the box. Use tubing cutters, so that the ends of the sleeve are cut square-shaped, as this is necessary to assure flat-mounting. The sleeve will maintain proper standoff, corresponding with the distance from the edge of the box to the wall surface, and will insure grounding continuity between the box and the device, too. Now the switch or receptacle can be firmly mounted, and the trim plate, when reinstalled, will fit properly and hug the wall.

In the same vein, sometimes wall finishers slip up and cut an opening too large in drywall material, so that the trim plate will not cover the entire opening. In walls of combustible materials, gaps between box and wall must be filled in with patching plaster or compound. Article 370 of the Code, under "Repairing Plaster," notes that no gaps or open spaces are to be permitted in the wall, unless it is of concrete, tile, or other noncombustible material. In either case, oversized trim plates are available to cover patched areas or rough edges of adjacent finished surfaces. These jumbo covers can be had in single-gang and multi-gang types, for switches or receptacles. They may not be stocked or readily available everywhere, but they can be ordered.

NO HOT WATER

Let us now examine possible causes of failure of operation in an electric water heater. The first place to look is in the panelboard, at the fuse or circuit breaker, to see if the fuse has blown or the breaker has tripped. As water heaters age, the heating elements, through deterioration, will sometimes develop current leakage to ground through the water in the tank before the element burns out. When that occurs, normal power consumption of the element when it is energized, coupled with fault-current leakage to ground, will overload the branch circuit, causing the protective device in the panel to actuate. The rate of fault-current flow may be just enough to load the circuit partially so that when it is added to the additional current flow when the element is called upon to heat, the total amperage draw that is created is just below the setting of the fuse or breaker. For this reason, restoration of the circuit may hold good for varying lengths of time, making the frequency of blowing the fuse or tripping the breaker unpredictable.

In a situation such as this, the elements in the water heater need to be checked for short circuits to ground with an ohmmeter. The ohmmeter is self-powered with a flashlight battery and registers a completed circuit when its leads are touched together or connected to components that are shorted together themselves. First, de-energize the branch circuit that serves the water heater. Then, remove the casing covers that conceal the elements. Disconnect the wires from the terminals of both elements, if two are present, and check each one for grounding. Touch one lead of the ohmmeter to the cold water pipe and the other to one and then the other screw terminal of the element. If the element is grounding out, the ohmmeter will register a circuit, indicating a path between the current-carrying part of the heating element and the cold water pipe, which is grounded. Either element exhibiting these characteristics must be replaced for safe, reliable operation.

If the overcurrent protective device in the panelboard is intact, the next step (for qualified personnel only) is to verify transmission of power to the terminals of the heating elements. De-energize the circuit that serves the water heater, as before, and remove the casing covers, exposing the elements. Now a voltmeter is used to detect the presence of voltage. If only one heating element exists, set the voltage scale selector of the voltmeter on 250 volts, and touch both leads to each of the wire termination points on the element. Now, while exercising caution, have someone momentarily energize the circuit. If voltage appeared on the terminals of the element, remove the wires, and, using the ohmmeter again, touch both leads to each of the terminals. If the ohmmeter does not register a completed circuit, the element is burned out and needs replacement. If voltage did not appear on the terminals of the element when the circuit was momentarily energized, touch the voltmeter's lead to the line side of the thermostat. If voltage is read there when the circuit is momentarily energized again, move the leads to the load side, and if voltage is not seen there when the circuit is momentarily energized, the thermostat needs to be replaced.

For dual-element water heaters, if the elements are interlocked so that only one can operate at a time, as is the case in many quick-recovery heaters, consideration needs to be given to the master thermostat and its transfer action when it is satisfied. If the tank is stone-cold, the top or master thermostat should have control and be placing voltage on the top element. If voltage is observed on the terminals of the top element, use the ohmmeter, as described previously, to check for element continuity. If no voltage is found at the element, check the thermostat to insure that a closed circuit goes through it.

A tank that is only partially filled with hot water, and hence apparently lacking the capacity to hold enough hot water to satisfy domestic loads, as it had done previously, points to trouble with the lower or slave element. With the top, or master, element satisfied, power is transferred to the lower thermostat, or slave element. The same tests can now be made for the lower element that were made for the top element.

The most common cause of loss of hot water or hot water tank capacity is associated with the elements in the heater. Trouble is infrequent, but occasionally elements do wear out. Proper tests can pinpoint the difficulty, and corrective action can be taken. Tests should be conducted by qualified personnel only, as exposure of energized electrical parts can be quite dangerous to the untrained. Replacement elements are usually readily available through most electrical supply outlets. The tank needs to be drained before removal of the old one, and, of course, it should be refilled and purged of air before being turned on again.

TROUBLESOME 3-WAY SWITCHES

Switches and receptacles do wear out after a while. Replacement is necessitated when the receptacle won't make good contact anymore, or when the switch will not control its outlet. Connection of the new device in like manner as the old one should be all that is required. Replacement is usually that simple.

A 3-way switch, however, may present a unique problem that first manifests itself as circuit outage whenever the switch is operated. A 3-way switch may serve satisfactorily for years until, one day, flashes may become noticeable when the toggle on the switch is operated. Finally the circuit overcurrent protective device will react, and the circuit will be dead, putting out lights in a portion of the home. To further confuse matters, the homeowner may attempt switch replacement, being careful to correctly reconnect the new switch exactly as the old one was done, and still find that operation of the switch causes the same difficulty as before.

To understand the problem, we need to recognize that existing 3-way switch wiring, performed prior to the mid–1940's, may be different in circuitry from the way it is required (by Code) to be performed today. The method used back then is often referred to as 3-way hot-to-hot which consisted of connecting a hot and neutral to each 3-way switch. As the switch was operated, it connected its common terminal to either hot or neutral. This being the case for both switches meant that in one instance the path for current through the bulb was hot-to-neutral in one direction one time, and hot-to-neutral in the reverse direction the next time.

Notice in Figure 68 that in the first circuit sketch, our path through the bulb from hot to neutral is from left to right. As shown in the second sketch, when the toggle on the right is operated, no current flows because the bulb is between two identical *hot* sources. No potential voltage difference exists. In

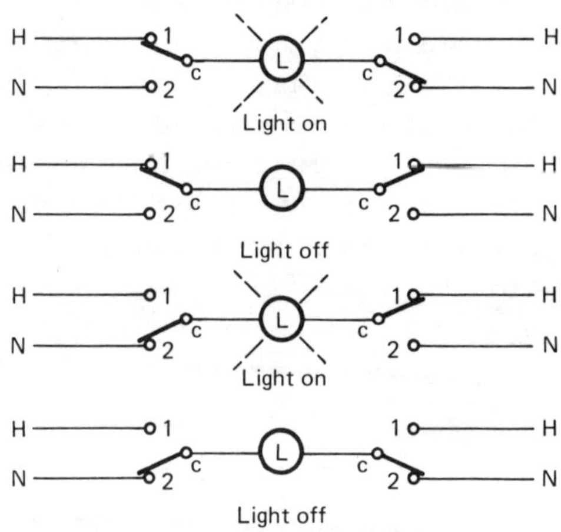

Figure 68: *Hot-to-Hot 3-Way Switch Wiring (Illegal to install as new work)*

the third sketch, when the toggle of the left switch is operated, the path is from hot on the right to neutral on the left, and the bulb burns. In the last sketch, when the toggle of the switch on the right is operated and the bulb is placed across two neutrals where no potential exists, the bulb does not burn.

The hot and neutral brought into each of the switch outlet boxes had to have been derived from the same branch circuit to insure that (1) de-energizing one branch circuit breaker or loosening one branch circuit fuse would positively kill power, and (2) no possibility could exist where, in the future during a service change, the two circuits might be connected so as to be fed from different legs within the panelboard. Such a situation would cause the bulb to burn normally in the first switch position but at double brilliance in the second switch position. Feeding each of the 3-way switches from different branch circuits fed from overcurrent devices in the panelboard connected to opposing hot legs would cause 120 volts to be impressed across the bulb in one instance and 240 volts to appear across the bulb in the other.

The method is illegal today. As we can see from the sketches, when the bulb is *out* due to switch positions that connect the bult to hot from both sides, the bulb is not *out* because no voltage is present but rather because no neutral is there to create a path. Consequently, a shock hazard exists to personnel who might work on this circuit, thinking it is safe because the bulb is not burning. For this reason, modern day Code requires that all switching take place in the ungrounded conductors, insuring that a lighting outlet when switched off will present no voltage potential to ground to constitute a shock hazard.

As we look at Figure 68 and the four sketches depicting various switch positions, it is apparent that each of the 3-way switches must exhibit *break-before-make* characteristics as the toggle is operated. Movement of the switch toggle must first definitely *break* contact with the neutral *before making* contact with the hot. If not, a momentary short circuit will exist, producing flashing and probably tripping the branch circuit breaker or blowing the fuse.

Therein lies the basis for the problem that arises when an old switch in a circuit like this begins to flash and cause circuit outage. The internals of the switch wear and loosen over the years. The common terminal wiper, as it moves away from one pole, drags that pole slightly because it is not firmly anchored anymore. Before it clearly breaks contact with one pole, the common terminal wiper "kisses" the contact of the opposite pole. As the toggle is operated, the common terminal wiper reaches the contact point of its new position just before it completely pulls away from the contact point of its old position. Replacement with a new 3-way switch may not cure the problem because the new switch may also not break cleanly with one pole, when the toggle is operated, before contacting the new pole position.

Three-way switches do, however, exhibit different characteristics in *break-before-make* qualities, perhaps not through defect, but because modern 3-way switch wiring techniques comply with Code requirements where switching takes place only in ungrounded conductors, and the *break-before-make* quality is not critical. Where one particular type of 3-way switch may not solve the problem, another will, if its *break-before-make* quality is quite good.

When replacing a 3-way switch in an old hot-to-hot circuit, you may examine the *break-before-make* quality of the new switch through the use of a meter which measures continuity. Most ohmmeters are satisfactory for this purpose. One ohmmeter lead should be attached to the common terminal of the switch, then the other lead to either of the two remaining terminals. Throw the toggle so you have a circuit. Then slowly operate the toggle to a position where the continuity is lost. Before moving the toggle further, move the lead from one switched terminal to the other. You should not see a circuit. Continued movement of the toggle should reclose the circuit for continuity only after *significant* toggle movement has been accomplished. If you see a circuit through the second switched terminal before further movement of the toggle is accomplished, you know that the 3-way switch does not have good *break-before-make* characteristics.

Switch construction sometimes varies among manufacturers. The old 3-way switches were designed for use on the hot-to-hot 3-way switch loops. Their contacts broke clean before setting up contact in the other leg. A little patience and understanding of the problem inherent with 3-way hot-to-hot switch loops will, however, get you back in business. Select a quiet switch with a click-free handle motion; one that does not exhibit a definite click. Then check it for proper switching continuity with the help of an ohmmeter and you will find reinstallation of the new 3-way switch will work. A switch which exhibits a *snap* when operated utilizes an off-center spring for snap-action and prevents our checking its break-before-make qualities.

If you suspect the possibility of 3-way hot-to-hot wiring, you can check by reading for the presence of a hot and neutral at each switch terminal. You can also examine the lighting outlet where you will most likely find two spare wires, one in each cable from each switch outlet. Only one wire was required to connect each of the fixture wires to different 3-way switch outlets, but BX and NM sheathed cable contains at least two insulated conductors, and one in each cable won't be connected.

In 3-way hot-to-hot switch circuits, good *break-before-make* switching characteristics are mandatory. The original 3-way switches installed had to exhibit a clean opening of the path from "c" to either "1" or "2" (see Figure 68) before establishing a path for current flow with the other. If this were not the case, a momentary short circuit would have existed as the common pole moved between its two contact points. And this is precisely what happens as the old 3-way switch ages and wears out. The switch internals when new were stationary, but as years pass and fatigue takes its toll, loosening of the anchoring of the electrical internals occurs, and the problem arises.

CONDENSATION IN CONDUIT

You would not normally expect deterioration of wiring where it is protected by steel conduit. Yet this same conduit which affords excellent mechanical protection and a good ground path for fault circuits can contribute to conductor failure over a period of years when low points in damp locations accumulate moisture. In years past, as the conduit system was installed, the electrician may have had to roll a U-bend into a length of conduit in order to turn up from the bottom of a panelboard. When this is done within conditioned spaces of a building, there is little danger of condensation within the conduit system. But sometimes, in areas such as basements which exhibit a higher degree than normal of relative humidity in the summertime, condensation can occur, resulting in circuit outages in later years.

Residential electrical systems of earlier vintages were candidates for "piped in" electrical wiring within the basement area. While later eras saw BX and NM sheathed cable pulled through floor joists visible from the basement, earlier methods made use of conduit hung to the underside of the joists with straps, with boxes and fittings which are still to be found in general use today where metal raceway systems are required.

Older homes which are wired with knob and tube quite often have wiring installed in conduit in the basement where circuits cross floor joints before rising up into the partitions. The conductors exist from the conduit through outlet heads which feature ceramic or other non-combustible bushing material. Continuation of the wiring from this point takes the form of the knob and tube system with insulators along side joists and studs and sleeves through joists and studs.

The conduit system in the basement will generally be of rigid steel pipe. Threaded ends connect to boxes, fittings and panels with locknuts and insulating bushings. The conduit system will last a lifetime and afford protection for the conductors within for many, many years. However, where the conduit may have been run off the bottom of a panelboard and then turned upward, a problem can sometimes arise within the loop formed by the upward turn. Over the years, condensation forming within the piping system will collect inside this loop, deteriorating the insulation covering the wiring within. In such cases, the wiring probably did not lend itself well to installations within damp locations anyway. As the combination of moisture and current flow sped the process of electrolysis between the steel conduit and the copper conductor, the wiring insulation broke down more rapidly. Thus, a problem situation arises when the circuit goes dead as a product of the conductor shorting to the conduit or the wire merely decomposing until a break occurs.

In responding to a trouble call such as this, the usual first steps are taken to diagnose the problem and locate the break or short. Reading each terminal within the panel to ground will tell us if the overcurrent device is passing power to the wire. If the overcurrent protective device will not reset, then breaking the wire joints in a systematic manner where they occur in boxes will assist in isolating the short to a specific segment of conduit. If the overcurrent protective device does reset, and outlets served by the circuit still remain dead, then reading for voltage at each joint within boxes will identify that section of pipe within which the wire is open.

When the first downward-then-upward loop off the panel is identified as being the pipe within which the trouble appears to be, replacement of the wire within this first run is in order. De-energize the circuit, and break the joints in the first box where joints are located. Pulling the old wiring from the pipe is the first order of business. When one of the conductors is broken, the other piece will have to be pulled back from the other direction. While the wiring is out of the conduit, drill two 3/16-inch holes in the underside of the conduit where it loops upward. This will promote drainage and insure that collecting condensation has a route of escape. If the conduit is also heavily rusted on the outside and all evidence points to serious corrosion of the pipe, replacement of the conduit run is in order, too.

Select new conductors suitable for installation in damp locations, such as wiring insulated with THWN type insulation, a moisture resistant thermoplastic. Insert a fishtape into one end of the conduit until its bullnose appears out of the other end. Set up the cartons of wire so they will self-feed from center-punched holes in the boxes, or if the wire is on small reels, arrange them to roll on a piece of ½-inch EMT as the wire is pulled off. Attach each conductor to be pulled in to the eye or loop of the fishtape by removing about 3 inches of insulation from each wire. Allow half the length of the exposed wire to be folded over through the eye or loop of the fishtape. Flatten the bends, so the wire lies close to itself after being passed through the fishtape loop. Then use electricians tape to cover the fishtape loop and folded over wire, being sure the tape is tightly pulled as it is applied to closely gather the wires into a tight little bundle. This prevents wire ends from snagging or pulling loose as the wires are pulled in. A second person can be used quite effectively to encourage the trailing wires to enter the opening of the pipe into which they are being pulled. Pull the fishtape from the pipe, and the new wiring will follow until you have the new conductors extending from the pipe, and ready to be reconnected. Remaking the joints in the box, and re-terminating the wires within the panelboard will restore the circuit and electrical service to the previously blacked out area of the home.

6

HOME APPLIANCES

AUTOMATIC CLOTHES WASHER

Today's automatic clothes washer is a necessary appliance in the home. So prevalent is its use, that the National Electric Code® recognizes its presence in Article 220 in instructions that require a laundry circuit load at 1,500 watts to be included in residential housepower calculations. And in Article 210, a special appliance receptacle for the washer is called for within 6 feet of the clothes washer water connections. Homes constructed today must automatically allow for use of the washer in their electrical construction. A home that does not feature a special-purpose washer circuit and receptacle will not be approved by the electrical inspectors, and for good reason, as shortchanging the home through omission of electrical connections for the clothes washer indicates extreme shortsightedness toward the homeowner's needs.

The clothes washer employs a reversible motor to permit transmission of motive force for agitation during washing and rinsing in one direction, and spinning the clothes tub basket for separation of water from the clothes through centrifugal force when the motor is run in the opposite direction. At the same time that spinning occurs, the tub pump is engaged in pumping out dirty wash or rinse water from the tub. Pump discharge is out through the drain hose and into a standpipe or service sink for disposal into the house drains. Spinning of the clothes tub basket approaches rather high revolutions per minute in some models of washers, so leveling during installation is necessary.

Washer construction features a tub within a tub. The inner tub is referred to as the basket. This is where the clothes are held. The basket is perforated to permit free interchange of water between it and the outer tub. During agitation, the basket may or may not oscillate slightly as the agitator assembly is oscillated in the opposite direction, depending on the model you have. Or the agitator assembly may pulse vertically instead of oscillating. During spinning, the basket whirls, and water escapes from the clothes and basket through the basket penetrations into the outer tub, where the pump suction in the base of the tub picks up the water for discharge.

Filling of water is accomplished by energizing electric water valves that are opened on signal from the sequence control mechanism to admit water. Varying water pressures that create different fill rates are compensated for with the use of a tub level switch to sense when the water within the tub has reached the desired level. Actuation of wash cycle is held off until the proper fill is achieved. Both hot and cold water connections are piped through electric solenoid valves, so admission of hot and cold water can be controlled for hot, warm, or cold temperatures in the tub.

Most washers today utilize a door safety switch to prevent machine operation when the door is open and to shut the washing machine down should the door be opened during its cycle. Manufacturers' internal wiring for the washer vary widely, but we will look at a simple schematic of how the electrical wiring might be accomplished in a basic washer.

Although we have mentioned some of the controls in the foregoing discussion, I believe that summarizing the various components that control the washer might be in order before looking at the washer's wiring diagram. We list the components, with their functions beside them, in an attempt to spotlight the control equipment found in the washer.

Sequence timer	Automatically steps the washer through cyclical functions, effects automatic shutdown, and permits manual mode selection.
Temperature selector	Indexes the machine to wash clothes in hot, warm, or cold water.

Load size selector	Establishes fill level cutoff points to correspond with amount of clothing to be washed.
Lid switch	Prevents machine operation until top is closed. (On some models, filling can take place with the top open, but agitation and spinning are locked out.)
Water solenoids	Open in response to electrical signal to admit hot or cold water to tub.
Water level switch	Monitors fill level of water in tub, and closes solenoids when desired level is reached.
Motor relay	Controlled by sequence timer to close its load-rated contacts to energize washing machine motor.

In Figure 69, the various control components are shown as they might appear in the schematic diagram installed on the washing machine. The lid

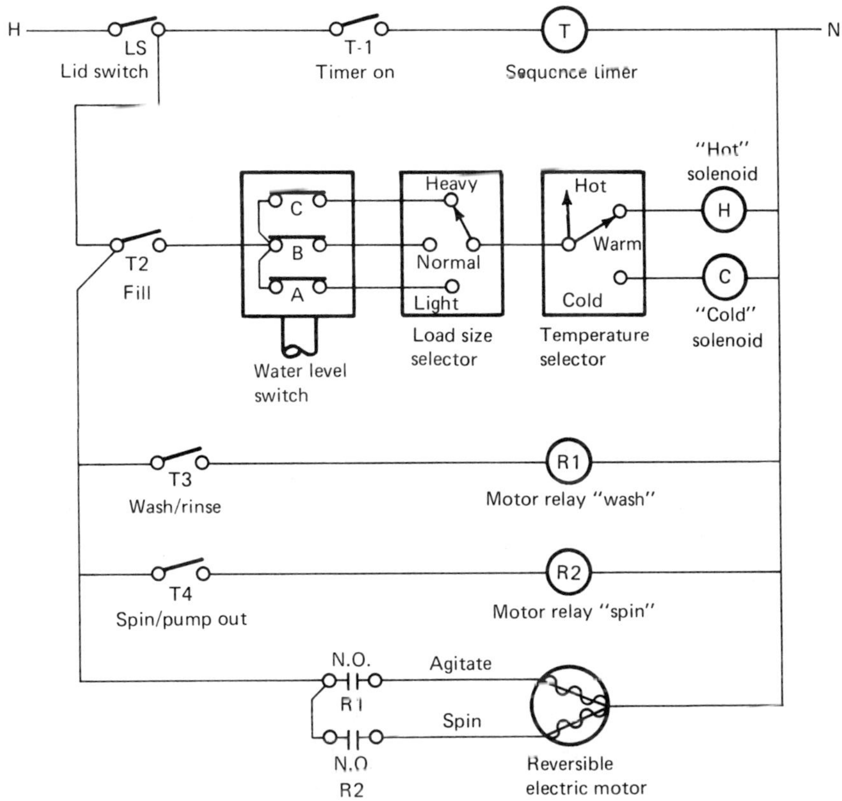

Figure 69. *Simplified Control Wiring Schematic for Basic Washing Machine*

switch is connected into the control circuit before all other equipment in the model pictured, so as to de-energize all operations anytime the lid is raised. Manual positioning of the sequence timer selector dial to "start" closes timer cam contact T1 to power the timer to keep it running until the end of the wash cycle, at which time the timer opens T1 to shut itself down.

Timer cam contact T2 is then closed to permit the washer to fill with water. A 3-pole water level switch that senses the amount of water in the tub opens its contacts, A, B, and C, in that order, on a rise in tub water level. The manual load size selector control switch located on the front of the backsplash panel permits selection of fill level to match the amount of clothes to be washed. In the interest of conservation, only the amount of water required for the load to be done is introduced into the machine. The circuit is completed to the water solenoids until the appropriate water level switch contact opens in accordance with the setting on the load size selector. The temperature selector manual adjustment switch is also located on the front of the backsplash, to allow selection of wash cycle water temperature. The double-pole rotary switch makes contact with the hot water solenoid when set for "hot." In the "warm" setting, both hot and cold water solenoids are energized; and in the "cold" position, just the cold water solenoid is allowed to open.

Timer cam contacts T3 and T4 are energized alternately, to cause the motor to run in first one direction and then the other, depending on whether agitation or spinning is required. When motor relay R1 is pulled in, its normally open contact closes to energize the appropriate motor windings so that the motor runs in the direction that engages the transmission components necessary to initiate agitation. When R2 is pulled in, its normally open contact closes to run the motor in the reverse direction, which drives the tub in a revolving manner and operates the pump impeller for pumping out the tub water.

In the event that the washer won't operate, make some basic checks. Is it plugged in? Is the water turned on? Have you closed the lid? Did you position the sequence timer control to start the machine? If these items check out, be sure to check the house power panel for a blown fuse or tripped breaker. This condition can arise if the washer receptacle is not on a circuit by itself. A separate 20-ampere 115-volt circuit and 20-ampere receptacle should be provided to serve just the washing machine, and no other loads.

If the washer won't fill, and the water is turned on, suspect the water solenoids. An ohmmeter can be used to check the continuity of the coils in the solenoids and detect an open circuit, which, if it does exist, is indicative of a need for solenoid coil replacement. Sometimes, strainers built into the ports of the solenoid valves will stop up from calcium deposits in the incoming water that collect on the strainer screen. It may be that cleaning the screen will restore normal fill rate.

ELECTRIC CLOTHES DRYER

Those wet, rainy days would have housewives at their mercy, were it not for the clothes dryer. Despite the weather, clean laundry, all dried and ready to fold, can emerge from the laundry area. The dryer has become a very commonplace item around the home, and the National Electrical Code® recognizes this by mentioning it in Article 220. But when the dryer doesn't dry, hubby hears about it, and swift action must follow. The laundry cycle depends on it.

Dryers should be vented to the outside. Hot, moisture-laden air is discharged from the unit's operation, and introducing it into the basement or under the house or into the utility area is a mistake. The unnecessary moisture adds to the humidity in the summertime and fogs windows near it in the wintertime. To accomplish venting, flexible hose made for the purpose, can easily be extended from the dryer discharge through a wall or window opening, using a weathercap and backdraft damper assembly, available in most hardware stores.

As an important member on the home operational scene, the dryer deserves some attention in this text. Repairs to get it going again are frequently not all that hard to accomplish, and some insight into dryer function and control might be beneficial in making a diagnosis, making repairs, or summoning aid. We shall now take a look at the "electrics" of the dryer performance, and see what happens when the dryer goes to work.

In the simplified wiring schematic of Figure 70, we note that the electric clothes dryer needs both 115 and 230 volts in its operation. The heating element uses 230 volts, but the tumbler motor requires 115 volts. This highlights the logic behind the requirements of many localities that electric clothes dryers be wired with 3-conductor cable that offers an insulated neutral, since in normal operation the neutral is a current-carrying conductor.

The schematic is elementary, and different manufacturers may have widely varying arrangements of components and somewhat differing techniques but the diagram will serve to demonstrate the basic operation of a clothes dryer. The following description will give us some insight into normal dryer operation; then we discuss maladies that sometimes arise and steps toward taking corrective action.

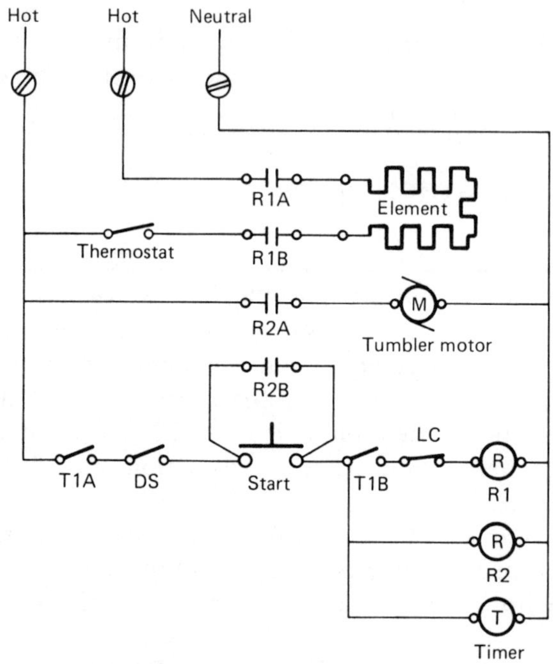

Figure 70: *Simplified Schematic of Electric Clothes Dryer*

Manual rotation of the master control timer selection dial to index the unit for a new drying cycle closes cam-operated contact points T1A and T1B. Closing the door to the dryer engages the door switch, setting up our circuit to the start button. Then the start button, normally open, is momentarily depressed, energizing heating element control relay R1, which closes its contacts R1A and R1B to complete the 230-volt power circuit to the heating element. Normally closed element limit control LC permits relay R1 to remain energized as long as sufficient air flow occurs across the heating element and element temperatures do not rise to unsafe levels. Should that happen, LC would open the circuit to R1 and cause it to drop out. LC is a thermostat located in the housing around the heating element. If R1 drops out, its (normally open) contacts open, interrupting operation of the heating element.

At the same time that R1 is energized, R2 is also energized, and one of its contacts, R2A, puts 115-volt power to the tumbler motor, while its other contact, R2B, sets up a holding circuit around the normally open, spring-loaded start button. Additionally, the master control timer is energized in order that the timed cycle may commence toward its completion, whereupon cam contacts T1A and T1B open, shutting down the dryer completely. The tumbler motor generally direct-drives the blower as well, so that the one motor accomplishes both drum rotation and air movement. Heat level controls, permitting selection of intensity of heat in some dryers, are not shown, but such selection could be accomplished merely by locking out part of a split-wired heating element. A thermostat controls heat temperature.

Were the master control indexed for fluffing only, or for drum rotation without heat, timer cam-operated contact T1B would remain open, and heat control relay R1 would not be energized. Opening the door to the dryer during operation interrupts the control circuit, drops out relay R2, and makes momentary depression of the start button necessary for resumption of operation. This is a safety feature to insure machine shutdown should anyone attempt to reach into the dryer while it is in motion, accomplished by door switch DS.

Notice in the diagram that the controls make use of the 115-volt circuit, as well as the blower-tumbler motor. The neutral is continuously conducting current during dryer operation, although not as much, and not at as high a voltage, as the wiring going to the heating element through relay contacts R1A and R1B.

Timer operation extinguishes itself when cam-operated contact T1A opens at the end of a drying cycle. This permits the dryer to attain automatic shutdown at the end of a cycle without human intervention. Complete shutdown may not occur on some models until a predetermined number of audible alerts have been sounded at timed intervals while the clothes are fluffed. That way, the dryer notifies us upon completion of the drying cycle and invites us to come remove the clothes and fold them before they become wrinkled from lying in a pile in the drum of the dryer. In these models, T1A is the last contact to open and disable the timer motor itself to prevent further recurrence of tumbler motor operation. Once T1A opens, further dryer operations can be had only through manual rotation of the master control timer dial to set up a new drying cycle.

What happens to an electric dryer when it doesn't work? Several different situations contribute to various types of problems, and we will now look at some typical troubles that can arise from years of usage, necessitating repairs.

The dryer will not dry the clothes. Upon examination, you see that tumbler rotation is occurring, but no heat is being imparted to the flowing air into the dryer drum. Look first at the fuses. With a voltmeter, you can check whether there is 230 volts on the loadside terminals of the safety switch or pullout fuse holder. On these terminals, 230 volts must be present for the dryer to heat. Do not read for the presence of 115 volts to ground on each terminal individually. One fuse could be burned out, and voltage from the good fuse would feed back to the loadside terminal of the bad fuse, giving a reading of 115 volts to ground on both terminals. Remember that with relay R1 energized, which could well happen when the leg feeding our 115-volt control circuit is intact, "hot" would travel through the contacts of R1 and the heating element, giving us a reading of the voltage on the loadside terminal of the other fused leg. Throw the disconnect switch and remove the fuses. If 230 volts were not present across the loadside terminals when the disconnect was closed, use an ohmmeter to check out the continuity through each fuse individually. Lack of continuity through one of the fuses identifies your offender, and simple replacement should restore dryer operation. If 230 volts is present on the loadside terminals mentioned, as it should be, examine the dial setting to determine if the dryer is actually indexed to the drying cycle, and not the "air only" cycle.

With the dial properly set, and still no heat in the dryer drum, examination of the heating element is in order. They do burn out, once in a while, after years of operation. After unplugging the dryer, expose the terminals to the element and disconnect the wires. Then, check out circuit continuity through the element itself, using the ohmmeter. If no circuit is observed, remove the element, and examine it for breaks. You'll probably see that it has burned in two, and installation of a new one will cure the problem.

If the element is okay, check the internal wiring carefully for breaks or signs of damage. Then, suspect the heat control relay, R1. It may not be picking up when voltage is impressed across it. Remove the wires from the holding coil of R1, and, using the ohmmeter, check through the coil for continuity. You should see a circuit through the coil of the relay, and if you do not, the coil of the relay is burned out. Relay replacement is in order, or at least coil replacement.

Should the wiring and relay check out okay, suspect the limit control thermostat, usually located in the side of the element housing, through which the draft is induced by the blower. Again, using the ohmmeter, remove the wires from the limit control and check through it for evidence of a closed contact. Here again, you want to see a completed circuit through the limit control. If you do not, limit control replacement is the ticket.

Assuming our findings with the limit control do not indicate trouble with it, we must then suspect cam-operated contact T1B of the master control. We can excuse cam contact T1A, the door switch, and relay R2, because they lie in series in a common circuit up to the point where the circuit splits just before T1B. Check circuit continuity with the ohmmeter through the heating contact T1B. If an open is found, replacement of the master control appears in order.

This methodical, detective-like troubleshooting approach will reveal the difficulty with the dryer heating, and guide us toward making the correct repairs to get it to heat again.

Suppose the dryer does not tumble the clothes. Warm air is blowing through, but the clothes drum does not revolve. With a dryer that utilizes a common motor for air flow and drum revolutions, we may discover that the drive belt for the drum is broken. This is frequently a long, slender belt that encircles the drum and connects to a motor pulley through an idler pulley. Usually, a diagram is posted inside the dryer in the vicinity of the belt to aid in belt replacement. Copy the make and model number of your dryer to show to your appliance parts distributor for purchase of a replacement.

Like many units, your dryer may best lend itself to belt replacement through front access. The top may be raised by releasing two spring-type clips and lifting, while the rear portion of the top deck pivots on hinges. This will expose the drum and reveal screws that can be loosened to permit the front panel to be removed. While you're in there, clean away all lint and dust that you might see. Invariably, a significant amount collects in the unit during its operation over a period of years.

ELECTRIC RANGE

One of the largest appliances in the home, in terms of connected load, is the electric range in the kitchen. The total connected load may be on the order of 12,000 watts, but recall from our Code residential load calculation example that we were required to consider only 8,000 watts in our deliberations regarding service equipment. Cooking is accomplished at modulated temperatures, not on a range where every burner is operating at its highest setting. For this reason, Article 210 of the Code permits us to size our range branch circuit conductors as small as #8, but cautions us that overcurrent protection shall not be greater than the capacity of the branch circuit. So if we use #8 copper cable, we must provide overcurrent protection of 40 amperes, in accordance with the conductor capacity tables in Code Article 310. With a 12,000-watt range connected at 230 volts, you can see that the amperage from branch circuit conductors and the overcurrent protection level is significantly less than the 52-ampere load that the range represents (12,000 watts ÷ 230 volts). And yet this 40-ampere circuit will serve without any difficulty for many, many years, attesting to the fact that normal range use does not call upon total range capacity. Ranges feature a 50-ampere cord and plug, and a 50-ampere receptacle must also be provided. The receptacle offers a 3-wire circuit, capable of furnishing 115 or 230 volts to different burners simultaneously. This allows modulation of burner output through the control switch, which may offer several distinct settings, or an infinite setting, anywhere from off to full-on.

Surface-unit temperature control can be effected through distinct-mode switching, which varies power connections to the burner in accordance with setting selected, or through thermal modulation in the infinite switch, which proportions the cyclical on-time of the burner upward as higher settings are selected. In distinct-mode switching, the heating elements within the burner unit are stepped through graduating voltage connections as the selector is positioned from "off" to "high." In a common application, the coiled heating element that constitutes the burner is designed for 3-wire connection, with one of the connection points being the midpoint of a continuous path from the outer end of one of the sections of the coil to the other section. In

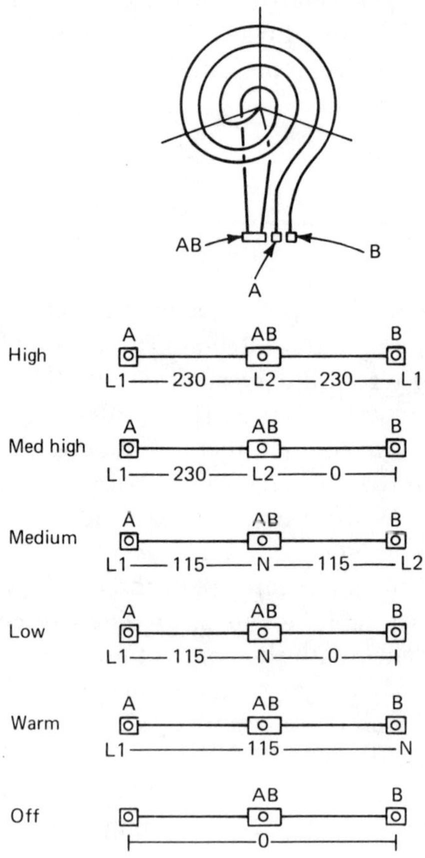

Figure 71: *Switching Connections for Distinct-mode Range-top Burner*

the type of burner unit under discussion, it may be observed that two distinct elements appear to be scrolled together, one interwoven or intertwined with the other, as it were, with a total of four leads trailing back under the cooking surface to their connection points. Two of these leads are commoned up, or terminated together at the junction with the range wiring. This sets up a midpoint tap in the total length of both heating element sections taken together, making it possible to apply different voltages to different sections at different times, subject to heating levels required and switch positions. In Figure 71, which illustrates distinct-mode switching, we see first the familiar scroll pattern of the range-top burner, and then the application of varying voltages at distinct heat levels, represented by markings that guide the way to switch selections that the homemaker makes use of to regulate cooking temperatures.

In the upper part of the illustration, we see the four leads identified as A, B, and AB, the common tie between the two element sections. The three wires extended to them from the control switch are alternately connected to impress varying voltages, as shown in the lower part of the illustration. In this manner, heat output from the burner is held to predetermined levels at each setting. "L1" and "L2" stand for the two out-of-phase 115-volt-to-ground voltages present at the range receptacle, between which there is a voltage potential of 230 volts. "N" stands for the neutral in our wiring system, which is also present at the range receptacle.

In the "off" position, no voltage at all is impressed across the heating elements. At "Warm," 115 volts is connected across both sections of the burner element, raising the burner temperature to a modest level, designed to maintain food temperatures at serving level only. When the selector switch is indexed to "Low," the 115 volts is reconnected to pass through just one portion of the burner element, causing half of the burner surface to become quite hot, while the other half is idle. This heat level might be selected to maintain cooking level temperatures just enough to permit extended slow cooking, particularly desirable in the preparation of some dishes. On "Medium," the second half of the burner is brought up to a hot temperature also, as 115 volts is impressed across it. This results in a uniformly hot burner, usually suitable for maintaining boiling temperatures. At "Medium high," half of the burner is reconnected so that 230 volts is impressed across it, while the other half is idle. The results in half the burner becoming cherry red and extremely hot, probably the desirable level for searing steak to seal in the juices before dropping the heat back slightly for pan frying. And, when the switch is moved to "High," 230 volts is impressed across both sections of the burner element, and the entire burner becomes uniformly cherry

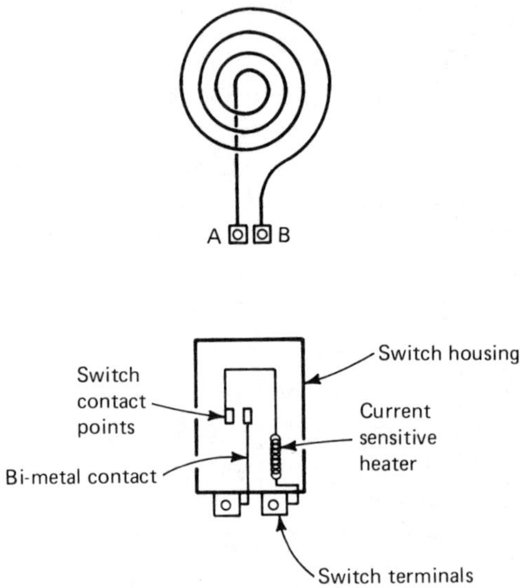

Figure 72: *Infinite Switch Control for Two-wire Burner*

red and ultrahot. This heat level is frequently selected when cooking is initiated, in order to raise the temperature rapidly to desirable cooking levels before adjusting the heat downward. It is the switch that does the job of alternating voltage connections to the burner. Then, these fixed voltages are maintained until the switch is readjusted for another heat selection. The control switch configuration might be rotary, push-button, or lever style.

In the infinite switch (see Figure 72), so-called because it theoretically permits an infinite number of settings in between "off" and "full on," the connection to the burner's single element is 230 volts whenever the switch closes its contact. However, current flow to the burner is routed through a heater within the switch itself, which reacts to the flow by becoming warm. As this happens, the bimetallic construction of the switch's movable contact begins to warp away from the stationary contact, subsequently opening the circuit, de-energizing the burner, and permitting the internal heater to cool. When the selector dial on the switch is moved to a higher setting, the movable contact is positioned closer to the stationary contact, so that a greater degree of warping of the bimetallic contact is now necessary to open the circuit to the burner. This increases the cyclical on-time of the burner, as the "closed" condition of the switch contact begins to become longer, while the "open" condition begins to become shorter. When the control switch is indexed to "High," the contact is mechanically held closed for full output of the burner without interruption. This switch differs from the distinct-mode, fixed-level switch in that the output of voltage from the switch to the burner is oscillated to effect the desired burner temperature. And each time the contact within the infinite switch closes, the full 230 volts is impressed across the burner element. The infinite switch is capable of offering closer control of burner temperature for a more accurate correspondence to the requirements of the food being prepared. The dial can be repositioned in extremely small increments to give us an extrememly wide span of adjustment positions and resultant heat levels.

Regardless of the type of burner element and switch control the range possesses, normal wear and tear do take their toll over the years, and switches and burner elements do have to be replaced. Appliance parts depots throughout the country offer excellent lines of replacement parts. For best service, you should always make a note of the brand and model number of the unit from which the old switch or burner came, and take the old one with you for matching of components at the counter. Usually, when burners burn out, physical evidence is apparent in the surface condition of the element itself. But this is not always the case. So an ohmmeter can be used to check through the element to see whether an "open" exists, a clear sign of a need for replacement. Checks can be made through the switch also. With malfunction occurring at just certain heat level settings in distinct mode switches, replacement is the appropriate remedy. Of course, be sure that the range is unplugged or the circuit disconnect is opened to insure personal safety before going into the wiring and electrical components of the range.

Oven control is effected through thermostatic regulation of actual oven temperatures. You can probably observe the sensing bulb of the oven thermostat mounted to one side of the oven interior. Its capillary extends to the oven temperature selector. Temperature within the oven is transmitted through the thermostat capillary to a temperature-actuated switch within the oven temperature selector. Oven heating elements are cycled on and off through thermostatic action to maintain selected temperature settings. Both the upper and lower elements within the oven are cycled on and off, but during baking, the upper element may be connected for 115 volts, while the lower one is connected for 230 volts. When energized during baking, the lower element can be seen to reach a cherry red condition, but the upper one usually does not. During broiling, however, the upper one is connected for 230 volts and does reach cherry red, while the lower one is idle. This give us an intense heat condition in the top of the oven, which is desirable during broiling. Generally, the thermostat setting will have to be moved to broil at the same time as the oven mode-selection switch. This is because on many models the thermostat remains in the circuit during broiling; hence, it must be set for maximum to forestall shutdown of the broiler due to oven temperature reaching its setpoint. Frequently, a pilot light, connected to light at the same time that the oven elements are energized, is provided to indicate when the oven is heating. This provides a visual check on oven operation and serves as a reminder to turn off the oven when we are through. A distinct click can generally be heard whenever the thermostat closes to cycle the oven elements on for maintenance of oven temperatures.

Self-cleaning ovens utilize exceptionally high temperatures within the oven to burn off collected cooking spills and reduce them to an ash. A special door interlock is usually provided that prohibits initiation of the cleaning cycle until the door is firmly closed and locked into position. Additionally, entry is barred during the cleaning period, as exposure to these searing temperatures could be quite hazardous. Special insulation prevents migration of these extreme temperatures to the surface of the appliance.

Continuous-cleaning ovens utilize a special interior that induces the breakdown and decomposition of spills during normal baking. This is supposed to preclude the necessity for the application of excessively high temperatures to clean out the oven. Continuous-cleaning refers to the fact that the design is intentionally and inherently self-cleansing during ordinary oven cooking activities.

Conductors used in the wiring of the range and oven have a special insulation directly suited to the purpose and rated at high temperatures to withstand the hot environment in which they must operate. Although the wiring does not enter the oven or come in contact with the heated surface units, it is nonetheless exposed to exceptionally hot temperatures due to its proximity to heating areas on the range, as it makes its way under the surface top of the range and up and down the back. This sounds a note of warning to anyone who might be faced with replacement of the internal wiring in a range. Be sure the proper type of insulation is selected. You will notice that the wire that you see after removing range backplates or lifting range surface tops features, perhaps, a fibrous, asbestos-looking kind of insulation. Selection of this particular type of wire by the manufacturer was no accident. Its special heat-resistant qualities are necessary to prevent insulation breakdown in this hot environment. Replacement of wiring must always be of the same type as that which is removed. Article 310 of the Code lists several types of wiring that may be suitable for the purpose. Types A and AA are rated for 392° F, and are listed as suitable for wiring within apparatus. It might be that the wire you observe is of this type. Any questions you have in this connection could probably be answered by your appliance parts distributor or your local electrical inspector. Do not use a wire that cannot withstand these temperatures, as its insulation will probably burn off, and subsequent contact with the range surface will produce serious arcing and circuit outage.

In recent years, the familiar convenience receptacle on the range has disappeared, and with good reason. Cords attached to small appliances that were previously connected to this receptacle frequently were draped across burner elements and either came in contact with the heated element, or, if the element was off, were not moved back out of the way when it was subsequently turned on. This resulted in severe cord damage and, consequently, an operational hazard to anyone who was working about the grounded surface of the range, because of possible contact with the exposed wiring of the cord. The range top is no place for the use of small, portable appliances. Their proper place is on the counters, connected to a receptacle served by a kitchen circuit. But these receptacles do exist on many of the older ranges, and when the receptacle ceases to work, you need to be aware of the possible presence of a special fuse or circuit breaker within the range itself to protect that receptacle outlet from overcurrent. Remember that the range circuit is protected for anywhere from 35 to 60 amperes, but this is inadequate protection for the convenience receptacle and any appliances that are connected to it. So a separate 15- or 20-ampere overcurrent device will be wired in ahead of the wiring to the receptacle to limit the current to normal outlet amperage.

Access to the range's built-in fuse or circuit breaker may be behind a removable access panel in the front, or concealed just behind the overlapping flange of the lower storage drawer, if your range is so equipped. In some cases, removal of the rear panel might be required. The homeowner's manual for your range probably locates the fuse or breaker for you; if you are in doubt, however, your appliance dealer who handles your brand of range should be willing to tell you what he knows about your model. If replacement of the fuse is in order, be sure you install one of the same size. Do not insert a 30-ampere fuse if a 15-ampere one is in there now. Overfusing could result in damage to the interior wiring to the receptacle outlet if too large a drain is imposed. The same precautions relevant to proper fusing in your house power panel need to be applied to replacement of fuses for the range receptacle.

If your range does feature this built-in receptacle, no doubt you have been using it successfully for many years. There is no reason for you not to continue to do so, assuming you continue to exercise ordinary caution and care in cord placement, being sure to keep use of small appliances limited, preferably, to times when the burners are not in use.

AUTOMATIC DISHWASHER

What a work saver today's electric dishwasher is, particularly on holidays, when many dishes are likely to have to be done. Now they can be stacked inside, the door closed, and the dishwasher commanded to wash and dry them, ready to be used again. First, dishes might be rinsed by the dishwasher, then washed, rinsed, and dryed. These functions are regulated by the dishwasher's built-in programmer, which is an electric timer motor operating cam-driven contacts in a predetermined sequence to initiate one event after the other. This programmer features a front-mounted dial, permitting manual advancing for start-up, with automatic shutdown at the completion of the dishwasher's cycle.

Dishwashers utilize a reversible motor that is mated to an impeller that is designed to circulate water for washing action when run in one direction, and to expell water for pumping out the dishwasher tub when run in the other direction. Pauses are placed in between reversal of direction to permit residual momentum to die out before energizing the motor for revolution in the opposite direction. An electric valve is opened at the appropriate times to introduce fresh water at the beginning of the rinse and wash cycles, and sometimes momentarily during pump-out for rinsing down the tub interior. This valve is a solenoid valve, for snap-acting, two-position closed-open operation, wherein energizing a coil raises a valve stem armature by induction against spring pressure to permit entry of water, and breaking of the coil's circuit allows the valve to spring-return to the closed position.

Operation of the dishwasher is prevented until the dishwasher door is closed and is interrupted if the door is opened during a wash cycle. A door switch, or door interlock, opens the circuit to the dishwasher anytime the door is ajar. This prevents injury to people that may result from operation with the door open and insures that the door is tightly sealed before initiation of a cycle can begin. The timer programmer is manually set to get the dishwasher going, and the four cam-operated timer contacts are opened and closed in a predetermined sequence to effect various events of the dishwasher cycle. The last movement of the timer opens the contact that shuts down the dishwasher until a new cycle is manually indexed again.

Dispensing of soap into the dishwasher tub at the proper time may be accomplished mechanically through timer-actuated cog tripping of an internal soap cup, which is filled as the machine is loaded. This cup is covered with a gravity cap that falls open when the cup is inverted. Thus, soap can be held out of the tub during prerinse and introduced during wash.

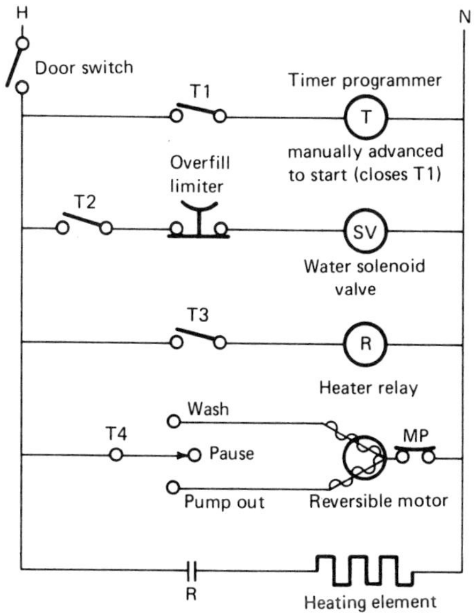

Figure 73: *Elementary Schematic Diagram of Basic Dishwasher*

In the elementary schematic diagram of the electrical circuitry of a basic dishwasher (Figure 73), we see timer programmer T, water solenoid valve SV, heater relay R, and the machine's reversible motor with internal motor protector MP. Cam-operated contact T2 would perhaps be closed initially, to introduce hot water into the tub of the machine. Then, T4 would make over to the wash contact, starting circulation of water within the machine for prerinse. Following a timed rinse, T4 would first move to "pause," then to the pump out contact, for removal of dirty prerinse water from the tub of our dishwasher. Then, T4 would return to "pause," while T2 was closed again for refill. T4 would then be indexed back to "wash," as our soap dish was inverted to spill soap into the machine. Water would circulate for washing during a timed period, after which T4 would come to "pause," then move to the pump out contact again for removal of dirty soapy water from the tub. Now comes rinsing, and T2 closes for more water, after which T4 makes contact with wash again to circulate fresh water for soap removal. Then pumping out would again take place, during which T2 might close momentarily to introduce water for tub rinse-down. After final pumping out, T4 would remain at the pause contact and T3 would close to energize the heating element relay for drying the dishes. Contact R would place voltage on the heating element for a timed period of drying. Finally, T3 would open, to stop drying action, and T1 would also open to de-energize the timer programmer for machine shutdown.

In the event that your dishwasher will not operate, the first place to look is in the panel box for a blown fuse or tripped circuit breaker. Dishwashers should be on a circuit by themselves, usually 20 amperes at 115 volts. Time-delay fusing should be utilized due to frequent starting of the machine's motor during a normal cycle. Inrush current during motor start-up may momentarily reach a level six times that of normal current, until motor revolution is established. Time-delay fuses permit this momentary current to pass without nuisance fuse blowing. Circuit breakers generally will not be affected by motor starting currents. Your dishwasher should be wired with 20-ampere #12 wire to match its 20-ampere overcurrent protection.

Most dishwashers make use of a resistance heating element that comes into play during drying of the dishes, after washing. This element could be 1,500 watts and, at 115 volts, would draw 13 amperes. A 15-ampere circuit would be loaded beyond 80% of its capacity, so the need for a 20-ampere circuit is apparent. Removable kick panels at the base of built-in dishwashers permit inspection of most control components and testing for presence of voltage during checkout and troubleshooting. The power cable is brought into a small junction box, where machine internal wiring is mated with circuit wiring via Wire Nuts®*, and grounding continuity is established. Grounding is very important, as fault-voltage accidentally introduced onto the surface of the appliance in the kitchen environment could give a shock to people working in the area of the sink.

Should the machine fail to fill, the water solenoid valve could be the culprit. An ohmmeter can be used to check for continuity through the coil of the water solenoid, and if there is an "open," replacement of the coil would be necessary.

If the machine fails to start at all, the door switch may prove faulty. Again, the ohmmeter can be connected to the wires that go to the door switch, and the door can be closed to see if a closed circuit occurs.

Should the dishwasher fail to pump out, first make sure the motor is responding properly, by reversing itself to run the impeller to expel water. If this is taking place, check for engagement of the impeller by motor shaft. Sometimes, impellers lose good mechanical contact with motor shafts through wear when mating surfaces strip free of one another. Most impellers are replaceable and available through appliance parts supply houses. As in all cases, your machine's make and model is essential information when seeking replacement parts.

If dishes don't dry, the heating element might be burned out; or the heater relay, if one is used, could be the problem. Our ohmmeter can be used here, too, to check for element and relay coil continuity. Some machines

*Wire Nut is the registered trademark of Ideal Industries, Inc.

feature an energy-saving selector switch that allows dishes to dry naturally from the residual heat of the hot water, rather than energize the heating element. Make sure that this selector switch is set to permit drying with the heating element. Avoid placing certain lightweight plastic utensils or tableware too close to the heating element in the bottom of the dishwasher. Check your owner's manual for precautions in this regard.

SUMP PUMPS

For basements in areas where high groundwater level occasionally forces water in through the walls or floor, special reservoirs, called sumps, are often constructed in a corner of the basement in an unobstrusive area. These sumps are well-like depressions, designed to catch seeping water below the floor level and pump it out through the house drainage system. Their value can be appreciated when you have goods or material stored in the basement, and you can't afford water damage. When sump pumps are used in conjunction with dehumidifiers, basements can be kept much dryer and more acceptable. Sump pumps can dispose of any excess water that may collect, while the dehumidifier condenses moisture out of the air to produce an internal atmosphere with lower relative humidity.

Installation of the pump itself would probably be by the plumbing contractor, since it is generally connected into the home's plumbing drainage. Wiring would then be by the electrical contractor and would consist of bringing a circuit over to serve the pump and connecting the pump through a disconnect to operate under control of its built-in level control. The unit would probably feature a float to react to the level of the water in the sump, and it would trip a switch to start operation of the pump on a rise in sump water level. When the pump would draw the water level down enough to lower the float and open the switch, pumping action would stop. Thus, the sump pump can react automatically to dispose of accumulating excess water.

Substantial amounts of water are likely to be removed by the sump pump during a very rainy period in an area where groundwater levels are frequently high. Note in Figure 74 that an under-floor drain tile is installed to collect water that would otherwise probably become a problem to the basement floor.

A special-purpose circuit should be extended from the panelboard to serve the pump unit. Its motor might be 1/3 hp, drawing 7.2 amperes and operating at 115 volts. A 15-ampere circuit here would be sufficient, protected

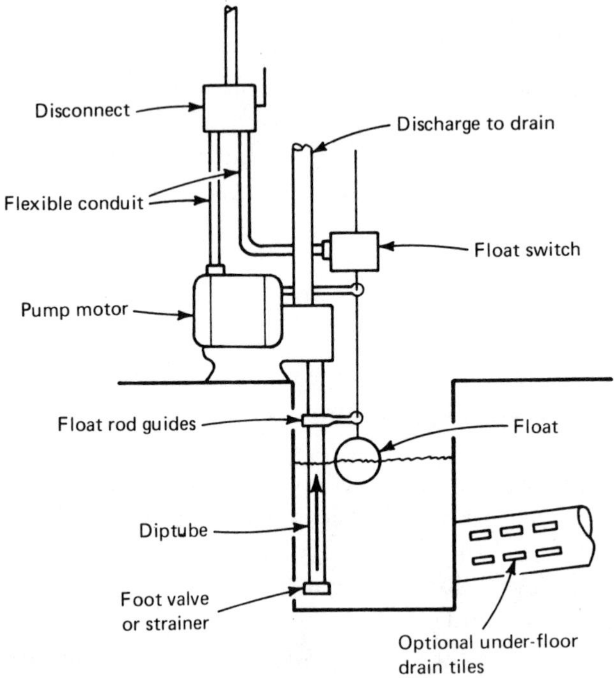

Figure 74: *Automatic Sump Pump*

with a time-delay type of overcurrent device to permit momentary start-up current to pass through without nuisance reaction. A disconnect switch should be located adjacent to the unit to facilitate convenient shut-down by anyone who might have to service the pump. Flexible conduit would extend from the disconnect to the pump motor and level-actuated switch.

The pump siphon or dip tube extends well into the sump to insure its being covered with water at all times. This prevents loss of prime and places the suction point well below the pump cutoff level. In the style of pump pictured here, a float rides the water level and imparts vertical movement to the level switch through the use of a float rod that rises up through the housing of the level switch to operate a set of contacts. As pictured here, our pump might very well be at the cut-in point, as the float is close to the lower rod guide. On a rise in water level, the switch contacts are closed, starting the motor.

There is a normal differential associated with automatically actuated contacts. In this case, it might be 3 or 4 inches of water level height. There must be enough differential to prevent short-cycling of the pump motor. It is desirable to minimize frequent starts and stops of the motor by allowing for 3 or 4 inches of draw down in level before the contacts open to stop the pump. Some models may even provide for 6 or 8 inches of draw down. The contacts of the level switch will usually be load-rated and operate the motor directly without the use of a relay or contactor. Residential models of sump pumps are generally not large enough to warrant considerations of intermediary starting relays or motor starters.

The dip tube should be equipped with a strainer to prevent the introduction of damaging solids into the pump intake, as they might severely injure the impeller. Or a combination foot valve with strainer might be utilized. The foot valve is actually a check valve that will permit flow up toward the pump, but prevent back-flow, which could cause the pump to lose its prime.

The sump should be covered, so that undesirable objects can't get in and people don't run the risk of inadvertently stepping into the opening. Covers with precut openings for pipeline and float rod penetration may be furnished by the pump manufacturer. The cover might be flat-hinged, allowing it to open into two half-circles, so as to permit embracing the pipe and rod. This facilitates easy removal for inspection and service to the float and suction line below in the water.

Joints in the electrical circuit to the motor could be made within the enclosure for the safety switch disconnect. The hot-in would be connected to the lineside terminal of the safety switch. Then, the conductor from the load side would be routed to the level switch. From there, another wire would leave the contact and run to the motor. From the motor, a neutral would be extended back up into the disconnect and jointed with the system neutral. In this manner, the disconnect and level switch are in series before the motor, so that the motor is isolated from any voltage anytime either the disconnect or level switch is opened. Insulation on wiring from the disconnect through the switch and on to the motor could be any color other than white, grey, or green; but the return conductor from the motor back into the disconnect to mate with the system neutral would have to be insulated white, to be ready identifiable as the grounded neutral.

DEHUMIDIFIERS

Your basement can become a much more desirable place through the regular use of a dehumidifier, designed to lower basement relative humidity through condensation and disposal of airborne moisture. Summer is usually the period when dampness in basements is most noticeable. Outside air temperatures may range into the nineties, and air at these temperatures can contain a significant amount of moisture. Basements, however, are partially subterranean, and cooler temperatures are to be found there, owing to the influence of ground temperatures against the perimeter walls. As outside air infiltrates the basement, its temperature is lowered, and its relative humidity is raised. This results in a very damp environment, which may show itself in the form of musty smells, warped paneling, damp fabrics, and maybe even mildew. The basement air needs to be subjected to dehumidification to lower its relative humidity and dry out the area. Steel tends to rust more in a damp environment. You can extend the life of your furnace, as well as other steel goods that may be stored in the basement, by drying out the basement.

Dehumidifiers are manufactured as self-contained units, designed to sit on the basement floor. When plugged into a convenient receptacle, they will operate under control of their built-in humidistat. By starting out with the humidistat set to the dryest setting, you can set it back in small increments until the dehumidifier shuts down, once an acceptable level of relative humidity has been attained. Then, this setting becomes the control point for continued automatic operation and maintenance of acceptable humidity levels.

The most popular type of dehumidifier uses a refrigerant cycle to condense objectionable moisture out of the basement atmosphere. The evaporator, or expansion coil, or cooling coil, is positioned first in the direction of air flow as air is introduced into the unit. The cold surfaces of the evaporator begin to "sweat" as air moves over it. Moisture gathers, and falls by gravity to collect in a reservoir, or is directed directly to a drain through a disposal hose. The air, as it leaves this coil, is quite cool, but is rewarmed as it passes over the condensing coil, before being discharged back into the room. The condensing coil serves the purpose of cooling the high pressure refrigerant gas off the compressor to cause the refrigerant to condense back into a liquid, so it may again be allowed to expand in the evaporator for further cooling of the incoming air. In a window-unit air conditioner or central-system air conditioner, the condensing coil is normally located outside of the house for the purpose of ejecting the unwanted heat in the refrigerant gas to the outside. But in the dehumidifier, the heat in the gas is ejected back into the room air to maintain relatively stable temperatures in the room after cooling by the evaporator for moisture removal. Thus, the heat taken from the air to force moisture condensation is returned to the air before it leaves the unit, but the moisture in the air is extracted from it, leaving the air drier and lower in relative humidity. A fan, integral to the workings of the dehumidifier, runs simultaneously with the compressor, both being controlled by the humidistat. Dehumidifiers are available in different sizes, so that unit capacity and cost can be matched to your needs, based on area served and amount of dampness to be dealt with.

Generally, the humidifier should be placed in service whenever the heating plant for the house is shut down for the warmer summer months. With the furnace going, moisture in the basement does not present quite the problem that it does when there is no heating taking place. However, in severe situations. it may be desirable to leave the humidifier in operation all year round.

If your unit does have a reservoir, means are probably provided to stop the unit in the event you neglect to empty the reservoir. This may be accomplished by a water pressure switch, which senses a pressure created by the rising water in a sampling tube, or by a level switch, which reacts to the weight of the reservoir as it fills with water. Many units include a pilot light on the front of the unit to signal that the unit is shut down from a too-high water level, rather than properly off from a satisfied condition.

Clean the coil that condenses moisture from the air periodically with a soft brush to remove any airborne lint and dust that has collected on it. If this dust is permitted to build up, it will act as an insulator and inhibit a dehumidifier from doing its job. Remember, the cold surfaces of the coil must come in contact with the air stream, so that anything that tends to insulate them from the air will result in a decrease in the amount of moisture that the coil can remove. If possible, use compressed air to blow the unit clean of dust and lint, paying particular attention to coil surfaces and coil fins. Cleaning of the dehumidifier is best done after it has been shut down for a while and the evaporator coil surface is dry. You may want to take it outside to take advantage of daylight to work by and allow dust to fall where it may.

PLANT GROW-LIGHT STAND

It's delightful to have fresh, growing plants in the bedroom or den. They add a touch of color and a sense of vitality to the atmosphere. When arranged on a lighted stand designed for the purpose, their maintenance can be quite easy, and the light from the stand can be quite beneficial to general illumination. Such a stand should be large enough to permit having relatively

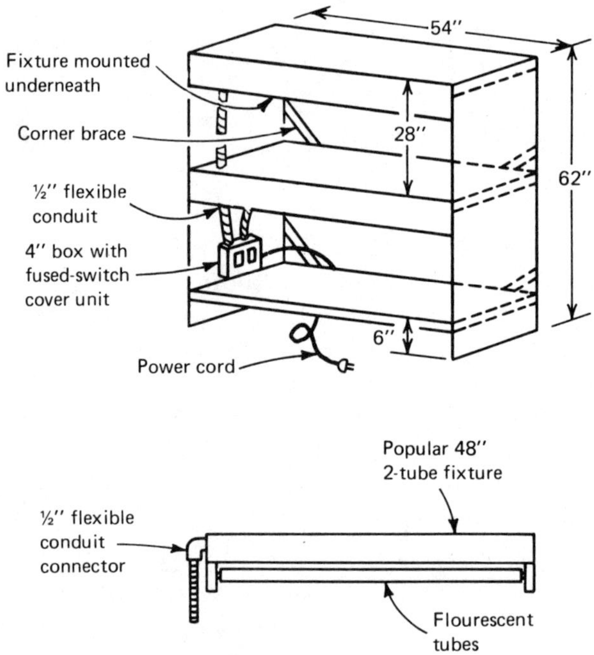

Figure 75: *Plant Grow-light Stand*

tall plants on it and to afford height for smaller plants to grow. When painted the color of the woodwork of the room in which it will be placed, the stand approaches having a built-in look when positioned against a wall, although it is quite free-standing. Fluorescent fixtures are mounted up against the underside of the shelf above, and books or whatnots can be placed on the top to help tie the stand into its surroundings.

Figure 75 shows a plant stand that can be built from three 10-foot lengths of 1- X 10-inch shelving board and one 10-foot length of 1- X 4-inch shelving board. The shelves and sides are cut from the 1- X 10-inch boards, while the front light guards are cut from a 1- X 4-inch one. Corner braces on the back side can be cut from 1/8- X 3/4-inch slotted aluminum flat stock. The 6-inch housekeeping height of the lowest shelf from the floor insures the ability to clean under the stand with wand-type vacuums or brooms. Of special consideration to this text is the electrical wiring associated with the construction of the stand. It will be designed to be self-contained, so that insertion and removal of a plug is all that is required to connect or disconnect the unit.

Two 4-foot fluorescent fixtures are selected to light the shelves of the stand. Cool white fluorescent tubes promote satisfactory plant growth, although special grow tubes are available. The color of the light from the special grow tubes may not be as desirable as the cool whites as an aid to general illumination, and that is the reason for the recommendation of the latter.

Each of the fixtures should be of the open type without lens, and mounted to the underside of the appropriate shelf with four #10 × ½-inch pan-head wood screws. Leave the fixtures open for wiring, and insert a 3/8-inch flexible conduit angle connector in the knockout located in the end of the fixture that is intended to be close to the switch. Set a 4-inch-square electrical box against the inside of the vertical end piece, about 1½ inches above the lower of the two lighted shelves. A 6-foot 16-gauge power cord should enter the box through a bushed 3/8-inch opening and be back-tied with an underwriter's knot to prevent pull strain from being imparted to the electrical connections within the box. The fusible unit cover should be fitted with a 6¼-ampere fusestat, which will permit us to use 16-gauge fixture conductors. Wiring from the fusible switch unit must be 3-wire, with the grounding conductor bonded to the box and to each fixture. Additionally, the circuit selected to serve the plant stand should be protected by a ground-fault interrupter for maximum protection of people, since the plant stand will be a wet area as a result of the presence of water from watering the plants, with subsequent probable accumulation in the saucers under the pots.

Flexible conduit-connectors in the ends of the fixtures will be angle-type connectors, whereas connectors in the top of the 4-inch box will be straight connectors. The former permits the flexible conduit to approach the fixture from a right angle. Note that our runs from the box rise straight to the underside of the fixture, so angle-type connectors are necessary for attachment to the end of the fixture.

Bore a hole through the upper lighted shelf close to the end to permit passage of one of our flexible conduit runs to the upper light. Measure the lengths of two pieces of 3/8-inch flexible conduit required, and cut them to reach from connector to connector. Then insert three #16 conductors in each piece with 24 inches of slack at the fixture end and 8 inches of slack at the box end. Pull the wires through the connectors, and fit the flex into the opening of the connectors and tighten. In each of the conduits, one conductor must be white, one green (the grounding conductor), and the other any color other than white, grey, or green. This last one will be the ungrounded or hot conductor, controlled by the switch in our fusible unit mounted on the 4-inch box. Use fixture wire approved for this purpose.

Make the appropriate joints, and bond the grounding conductor with approved grounding clips. This will establish continuous grounding integrity throughout our lighted plant stand. Incoming hot from the power cord will terminate first on the switch, so that it is first in line, and the wire will be circuited on through the fuse before being jointed to each of the two switched legs that go to the fixtures. White, of course, will be jointed to the white neutral in the power cord.

The connections are relatively simple, and the precautions are basic to good electrical work, designed to insure safe usage of the plant stand in an environment likely to be susceptible to occasional contact with water or damp earth. Much enjoyment can be derived from building the plant stand, and your safe enjoyment is assured through conformance to recognized safeguards.

Consider utilizing watering pans on the shelves to protect them from water damage that frequently accompanies arranging and watering the plants. These pans could be ordered from a metal fabricator, and might be constructed from aluminum that is perhaps 50 inches long and 9 inches wide, with 2-inch-high sides. Instruct your supplier that they should be watertight to prevent leakage. Then, gravel can be arranged throughout the pan to a depth of about an inch. Saucers can then be omitted from under the pots of the plants, and the pots can be placed directly on the gravel bed. Water that seeps through the bottom of the typical clay pot will then be dispersed throughout the gravel and allowed to evaporate naturally, adding nicely to the needed humidification requirements of the average home during the winter months, while heating is taking place. But the pans will have to be cleaned annually, for dirt and debris from potting and watering will collect in them, creating a muddy situation in the bottom of the pans if allowed to gone on for an indefinite period of time.

The final touch might be the addition of a timer to control the on and off periods of our plant stand's lights. This may be a simple plug-in type, into which the power cord to our stand can be inserted. Make sure the timer offers a grounding type of receptacle, so as to maintain our stand's grounding system with the grounding conductor of the ground-fault-protected circuit that serves it. Then, lights can be turned off periodically, in accordance with the plants' requirement for some period of darkness, which they would experience outside in their natural environment.

EMERGENCY LIGHTING

Occasionally, the power goes out due to the weather, and homes are plunged into darkness. This can be quite disconcerting, particularly to older people who may be experiencing some difficulty in getting about. Oftentimes, people resort to candles until power is restored, but to those who are finding their agility limited by advancing years, setting lit candles about the house might not be an altogether good idea.

Commercial buildings for years have been taking advantage of emergency lighting that automatically comes into play whenever power is lost. In department stores, for example, if the lights go out, true darkness may prevail, due to the absence of windows. So self-contained battery-powered units are employed to light strategic areas to insure public safety during egress.

The same practice can be introduced into the home, and it is sure to find favor among those whose activities within the home suffer in the event of sudden loss of lighting. The emergency lighting packs can be located anywhere and plugged into any convenient receptacle. As long as commercial power is available to the residence, a built-in relay is held in the open position to prevent the emergency lighting unit from coming on. Additionally, the battery that powers the lighting unit when power is lost is allowed to recharge, so that it is always ready to energize the emergency lighting. As soon as the electricity from the utility goes out, the monitoring relay closes to bring on the emergency lighting. Then, for as long as the power is off, the light will remain on, subject to the time limit on the battery's ability to provide standby service. The lighting in the emergency power pack might be 12 volts D.C., and the illumination provided more closely resembles an automobile headlamp than an ordinary interior home lighting fixture. But the light provided is sufficient to see how to get about and perform most necessary tasks.

Upon restoration of electricity from the local utility, the emergency lighting unit goes out, and the battery is permitted to undergo recharging for as long as necessary to restore it to the full charged state to serve again in future blackouts. This makes the unit quite self-contained and relatively maintenance-free. A test button is frequently provided on emergency lighting units to permit periodic testing to make sure they will light when called upon to do so.

Emergency lighting installation locations in commercial buildings are quite often high up on the walls, and their appearance within the store or office building is probably pretty much ignored. But in the home, location on a wall could be considered objectionable. Think, instead, of perhaps placing the units close to floor level, on the lower shelves of bookshelves, or on racks situated in unobtrusive areas. This way, the beams can be angled to provide light along regular access paths throughout the home from fairly concealed vantage points where the presence of the emergency lighting unit is generally unnoticeable.

7

EPILOGUE

Many of the basics examined in the preceding pages of this text carry over into electrical work in general, whether such work is residential, commercial, or industrial. The safeguards employed or methods utilized quite frequently find application whenever electrical work is under way, because they are basic to the practicioner and relate to fundamentals that are valid in every instance. The residential wiring specialist may be an individualist in the electrical trade in application techniques peculiar to home construction, but is a conformist when it comes to basics that govern his electrical practice. His working "bible" is the National Electrical Code®, and his consultants, when questions arise, are the electrical inspectors of the jurisdiction in which he is working. He takes pride in his work well done, and he relishes the approving nod of the local inspector upon examination of his work.

Whenever you perform electrical work, keep in mind the fundamentals that follow, as they are important to safe and efficient practice and are employed automatically by the journeyman electrician as he goes about making electrical installations.

Disconnect the electrical power before working on equipment

Even in receptacle replacement, it is easier and safer to remove the fuse or throw the breaker. Sure, it can be worked hot, but why go to the trouble? Your motions are faster and uninhibited when the circuit is dead and there is no danger of shock. Find the disconnect, and de-energize the circuit. It is safer and surer.

When equipment, such as appliances, must be gone into for troubleshooting in regard to the integrity of the circuit to the piece of equipment, de-energize the circuit first. Access to the termination point of the circuit in the piece of equipment is quicker and safer with the power off.

Sometimes turning power off is impossible, and panelboards and such must be entered hot. Doing so, however, should be limited to qualified people. When in doubt, don't. If you are uneasy about any segment of electrical work, stop and call for a professional.

Energize the circuit last

Consider an extension to a circuit within a home to serve a newly added lighting outlet. The wiring should be installed, all connections to the switch and fixture should be completed, and covers or plates should be in place before the circuit extenion is tied in. If it is tied in and energized prematurely, a hazard exists to the unsuspecting, particularly children, until the termination points or splices are properly covered. And in open energized work, such an accident could take place before you know it.

Ground metal raceway systems

Metal conduit is used in residential work where drops are required on basement or utility room walls to serve laundry receptacles or other surface-mounted convenience outlets. Additionally, it can be seen on the outside of homes in which it is necessary to extend circuits that would be inaccessible from any inside route. Though non-metallic-sheathed cable is generally fine when run within joists or studding, it must be given protection when it is to be installed exposed on inside or outside walls. Then metal conduit is used, and the cable may enter it in the ceiling joist space and be protected as it drops down the wall to reach the outlet box. Metal grounding clips, approved for the purpose, are utilized to bond the circuit grounding conductor to the outlet box. Then, through conduit-connectors approved for the purpose, the grounding continuity is preserved along the length of the metal raceway.

When circuits are to be extended outdoors, a junction box should be used, thereby permitting the indoor cable to enter the box through a knockout fitted with a cable connector and allowing the conduit to be connected to the box through a conduit-connector. Then, bonding of the junction box will ground the conduit. This does not, necessarily, remove the need for an equipment grounding conductor in the circuit through the conduit. It merely assures the mechanical and electrical integrity of the raceway itself.

Make splices only within enclosures

I'm sure we've all seen, at one time or another, one cable connected to another with wirenuts and tape, dangling in midair without benefit of a junction box for support and protection. This is an illegal method of splicing two cables together. The joints are unprotected, and quite often bear the weight of the cables. Such a joint is hazardous and can be quite dangerous, should the cable be snagged and pulled apart.

When necessary to connect one cable to another, a box of sufficient size must first be set in an area likely to remain accessible. Then, the cables are inserted into the box through approved connectors, which grip the overall cable and its jacket, affording relief from strain to the electrical joints within. A minimum of 6 inches of slack conductor length is left within the box, and joints are made using correctly sized wirenuts. The grounding conductors are properly bonded together, and a pigtail is bonded to the box itself.

When requirements dictate that a cable must be cut for tie-in of an additional cable to serve another area, it will sometimes be observed that there is insufficient slack in the cable itself to permit resplicing. In this case, two alternatives present themselves, short of extending new cable from the last outlet point, which may be impractical. Two junction boxes can be utilized, spaced far enough apart to permit 6 inches of cable to enter the box. Then, an additional length of short cable can be used to connect the two junction boxes. One of the boxes will then represent the tie-in point for the run to the new area.

Or a single enclosure, similar to a wiring trough or wireway, can be used, measuring perhaps 4 inches by 12 inches. Its length permits jumpers to be installed within to connect the butt ends of the severed cable. It may feature knockouts for installation of connectors, or a hole saw can be used to cut openings for the cable connectors.

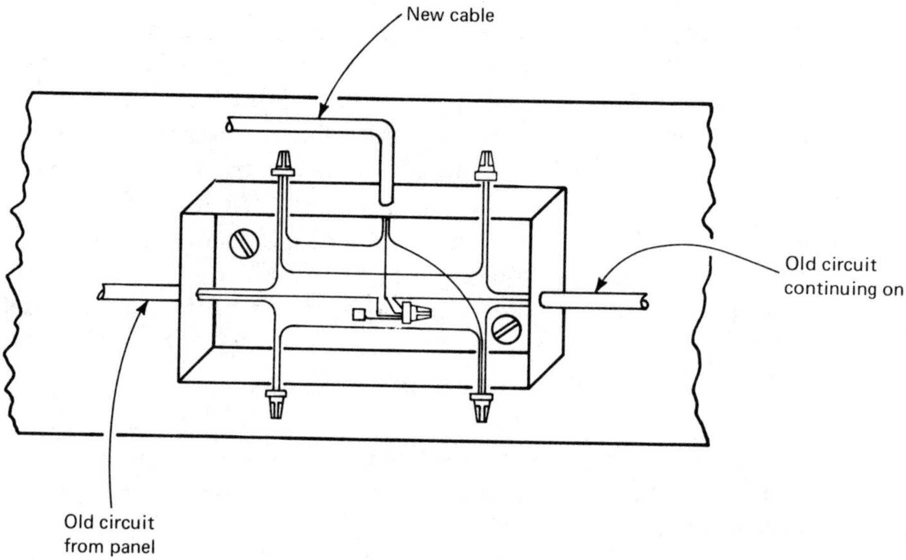

Figure 76: *In Line Splice To Existing Cable*

In either case, avoid the use of a small, single box, that limits space for splices and whose connectors clutch the cable so close to the cut ends that skinning, spreading and splicing can't be accomplished.

In Figure 76, we see the 4- by 12-inch wireway with its cover off, and splices made up. We will assume an attic location, so installation is against the side of a ceiling joist, just below the top edge. Were the attic to be floored, an access door would have to be left at this point to render the joists within the box accessible. It might be that the installation of the new cable was caused by the need for a pull-chain attic light. Here, the circuit serving the floor below is utilized through this splice to serve the attic light as well.

Terminate tightly

Loose connections heat up. Electrical connections that are not made up tightly allow arcing, as electricity attempts to flow across the voids left in poorly made terminations. Care is required in securing conductors to terminals and in securing one conductor to others, to insure that they are tightly bonded together to form a good current-carrying path with sound mechanical connection properties. This keeps resistance at the connection point to a minimum and provides for cool operating, trouble-free terminations and connections.

Connections that are poorly made and that exhibit heating characteristics will eventually deteriorate, with damage to conductor and terminal, evident as burning or discoloration of the metal, build-up of carbon deposits, and burning off of insulation, plus loss of temper in the metal of the wire; and these conditions merely compound the problem, and the connection goes from bad to worse. Of course, there is the danger of fire, in the event that current flow is high enough to generate a damaging temperature rise in ambient surroundings, or combustible structural members are heated to the ignition point.

Some electrical devices offer alternate methods of terminating wiring coming to the device. You may notice a choice of screw terminals or poke-home termination slots, allowing the installer to bind the wire under the screw terminal or simply push the bared end of the wire into a hole that incorporates an automatic catch, preventing pullout. Generally, the poke-home opening does not offer the square-inch surface area of contact that the screw binding post does, and the conscientious installer's preference would probably be for the screw terminal, particularly in cases where the device may be called upon to serve a load of some significance with respect to the capacity of the device itself.

But well made, tight terminations become an automatic habit of the professional electrician, because he recognizes the importance of maintaining the highest integrity in his electrical system. A 15-ampere circuit is only as good as the weakest current-carrying point in the circuit, and when partially loaded, the branch circuit overcurrent protective device will detect no difficulty in circuit operation, even though a poor joint may be reacting hazardously to current flow.

To protect against possible inaccessibility of joints in electrical systems, Article 370 of the Code requires that all connections in boxes be accessible and forbids the concealment of splices and taps by construction. So you can be sure that in your home there are no joints within the walls, and the cable is continuous and unbroken from outlet point to outlet point. With today's excellent products in the line of electrical installation equipment and devices, it is easy to make good connections and splices, and there is no reason inferior work cannot be avoided altogether. And the presence of the U-L symbol on items to select from helps insure high quality and acceptability.

Never install grounding-type receptacles that are not grounded

It's a temptation when called on to replace a receptacle to use a grounding-type model in place of a non-grounding type when that's all you happen to have. But don't do it. Instead, go get a non-grounding type, if there is no means for connecting the grounding lead within the enclosure. Article 210 of the Code clearly tells us that the grounding means must be present within the receptacle outlet box before a grounding-type receptacle can be installed. Installing a grounding receptacle without benefit of having the grounding terminal of the receptacle properly connected and grounded is a Code violation, and it represents a hazard to the unsuspecting user if the appliance or tool connected to the receptacle sustains an internal fault to the surface of the tool.

Apply for an electrical work permit

This is for your safety and those who come after you. Even a homeowner owes it to himself and subsequent owners to have his electrical work approved by local inspections officials. Remember, they exist for the benefit of the public, and subsequent owners of your house have the right to the approval by local authorities of work that you have done. Save your inspection ticket. This will show prospective buyers who may notice new electrical work that such work was approved by appointed inspections officials of the locality in which they anticipate purchasing a home.

Mount outlet boxes first

In new work, the route of branch circuit cable is better established when all boxes are in place. Much cable is saved, too, when cables are pulled with ample slack, stubbed into boxes, back-pulled, and stapled. Even in commercial work, boxes are roughed into walls as the walls go up, and later, conduit systems are extended to serve the various outlet boxes.

When in doubt, ask

If there is a section of the Code that you do not understand and that has not previously been explained to you, and you are preparing to do work to which that particular section of the Code pertains, don't assume an interpretation. The inspectors are happy to discuss the matter with you before the work is begun. Then, you can proceed, assured that your methods will find favor with the inspector at inspection time.

Special situations oftentimes give rise to questions of acceptability of methods to be employed in attempting to meet the Code. You may be uncertain as to how your inspector will look upon your implementation of the Code in any particular work situation. Don't assume. Pick up the phone and call him. You'll find out that he's interested in helping you do it right the first time. And you'll save time and money over having to redo the work should your assumption have been incorrect. The inspections officials are the experts on the electrical code in your area, and they are there to protect and assist the public and electrical contractors.

INDEX

A

B

D

J

K

L

M

N

O

P

R

S

T